Cambridge Tracts in Mathematics and Mathematical Physics

No. 47

CONVEXITY

CONVEXITY

BY

H. G. EGGLESTON
Professor of Mathematics in the
University of London

CAMBRIDGE
AT THE UNIVERSITY PRESS
1966

PUBLISHED BY
THE SYNDICS OF THE CAMBRIDGE UNIVERSITY PRESS

Bentley House, 200 Euston Road, London, N.W.1
American Branch: 32 East 57th Street, New York, N.Y. 10022
West African Office: P.M.B. 5181 Ibadan, Nigeria

First printed 1958
Reprinted 1963
Reprinted 1966

First printed in Great Britain at the University Press, Cambridge
Reprinted by offset-lithography by Lowe & Brydone (Printers) Ltd, London, N.W. 10

CONTENTS

PREFACE

Although convexity is used in many different branches of mathematics there is no easily available account dealing with the subject in a manner which combines generality with simplicity. My aim in writing this tract has been to provide a short introduction to this field of knowledge for the use of those starting research or for those working on other topics who feel the need to use and understand convexity.

In a short tract, on a subject such as this, it is difficult to decide both the level of generality to aim at and the exact parts of the subject to omit. On the one hand, to accommodate the needs of economists and others, it is desirable to have available results that refer to n-dimensional real Euclidean space; on the other hand, more general spaces present such diverse characteristics that they cannot be conveniently dealt with in a tract of this size. For this reason the containing space is taken to be n-dimensional real Euclidean space except in the last two chapters. As to the subjects omitted there is nothing on the geometry of numbers, packing or covering problems, differential geometry on convex surfaces, integral geometry or the analogy with complex convexity.

The tract falls naturally into three parts. The first and third chapters contain the basic properties of individual convex sets and functions. The second chapter is an illustration of the way in which the comparatively simple properties obtained in the first chapter can be applied. In the fourth and fifth chapters convexity is investigated more fully, the properties of classes of convex sets are developed and the effects of certain operations on these classes are studied. The last two chapters contain examples of results and techniques in the solution of particular problems.

The notes at the end of the tract contain brief indications of the sources of the material in the tract and of suitable papers or books for further reading. Other bibliographies will be found in the books referred to there, in particular those by Bonnesen and Fenchel, by Hadwiger and by Fejes Tóth.

My thanks are due to Dr F. Smithies, Fellow of St John's College, Cambridge for inviting me to write this tract and for reading the manuscript; and to Mr B. J. Birch, Fellow of Trinity College, Cambridge, for reading the proofs with a critical eye. Apart from correcting many minor mistakes both have made suggestions for improving the text that have been of great value.

H. G. E.

7 HAUXTON ROAD,
TRUMPINGTON,
CAMBRIDGE

27 August 1957

CHAPTER 1

GENERAL PROPERTIES OF CONVEX SETS

1. Preliminaries and notation

All the numbers that are used in this book are real numbers. Complex numbers are never used at any stage.

The sets of points with which we shall be concerned will all be subsets of a real n-dimensional Euclidean space. Points will be denoted by lower case clarendon letters and sets by German capitals, except that we use the lower case clarendon letter of a point for the set consisting of that single point. The frontier of a set $\mathfrak{X}$ we denote by $\mathrm{Fr}\,\mathfrak{X}$, its interior by $\mathfrak{X}^0$ and its closure by $\overline{\mathfrak{X}}$. In Chapters 6 and 7 we shall sometimes use capital letters to denote points. Numbers will be denoted by small Roman or Greek letters. n-Dimensional Euclidean space itself will be denoted by R^n. It is the class of all ordered sets of n real numbers $\mathbf{x}=(x_1, x_2, \ldots, x_n)$ made into a metric space by defining the distance between $\mathbf{x}$ and $\mathbf{y}$, where $\mathbf{y}$ is $(y_1, y_2, \ldots, y_n)$, to be

$$|\mathbf{x}-\mathbf{y}|=\left[\sum_{i=1}^{n}(x_i-y_i)^2\right]^{\frac{1}{2}}.$$

It is convenient to regard $\mathbf{x}$ as a vector and to define $\lambda\mathbf{x}$, $\mathbf{x}+\mathbf{y}$, $\mathbf{x}.\mathbf{y}$, $\mathbf{x}-\mathbf{y}$, by $\lambda\mathbf{x}=(\lambda x_1, \ldots, \lambda x_n)$, $\mathbf{x}+\mathbf{y}=(x_1+y_1, \ldots, x_n+y_n)$, $\mathbf{x}.\mathbf{y}=x_1y_1+\ldots+x_ny_n$, $\mathbf{x}-\mathbf{y}=\mathbf{x}+(-\mathbf{y})$ respectively where λ is a real number.

The origin in R^n is denoted by $\mathbf{O}$. The line joining the two distinct points $\mathbf{x}$ and $\mathbf{y}$ is denoted by $\mathbf{xy}$.

The symbol $\{\}$ will be used to indicate sets which satisfy conditions that will be stated explicitly inside the braces.

The empty set or the void set, regarded as a subset of R^n, will be denoted by ϕ.

The distance function has the following properties:

(i) $|\mathbf{x}-\mathbf{y}|\geqslant 0$ and $|\mathbf{x}-\mathbf{y}|=0$ if and only if $\mathbf{x}$ is $\mathbf{y}$.

(ii) $|\mathbf{x}-\mathbf{y}|+|\mathbf{y}-\mathbf{z}|\geqslant|\mathbf{x}-\mathbf{z}|$.

(iii) $|\mathbf{x}-\mathbf{y}|=|\mathbf{y}-\mathbf{x}|$.

(iv) $|\mathbf{x}-(\mathbf{y}+\mathbf{z})|=|(\mathbf{x}-\mathbf{y})-\mathbf{z}|$.

(v) $|\lambda\mathbf{x}-\lambda\mathbf{y}|=\lambda|\mathbf{x}-\mathbf{y}|$ if $\lambda\geqslant 0$.

The sphere whose centre is $\mathbf{x}$ and whose radius is r is denoted by $\mathfrak{S}(\mathbf{x},r)=\{\mathbf{y}:|\mathbf{x}-\mathbf{y}|<r\}$. A subset $\mathfrak{X}$ of R^n is bounded if there exists a sphere $\mathfrak{S}(\mathbf{x},r)$ such that $\mathfrak{X}\subset\mathfrak{S}(\mathbf{x},r)$. An important property of such a set is that for any $r'>0$ it contains at most a finite number of disjoint spheres of radius r'.

The distance of a point $\mathbf{x}$ from a set $\mathfrak{Y}$ is defined by

$$\rho(\mathbf{x},\mathfrak{Y})=\inf\{|\mathbf{x}-\mathbf{y}|:\mathbf{y}\in\mathfrak{Y}\}.$$

A set of the form $\mathfrak{U}(\mathfrak{Y},r)=\{\mathbf{x}:\rho(\mathbf{x},\mathfrak{Y})<r\}$ where $r>0$ is said to be a neighbourhood of $\mathfrak{Y}$.

We shall use the sign $+$ both for the addition of numbers and for the vector addition of sets (cf. Chapter 1, § 2, Chapter 5), but not for unions of sets. However, for differences of sets, the intersection-complement notation is cumbersome, and so we shall use the notation $\mathfrak{X}\dot{-}\mathfrak{Y}$ to indicate the set of points which belong to $\mathfrak{X}$ and not to $\mathfrak{Y}$.

The volume of a set in R^n is its n-dimensional Lebesgue measure. We denote the volume of the set $\mathfrak{X}$ by $V(\mathfrak{X})$ or by $V_n(\mathfrak{X})$ if we wish to emphasize the dimension of the space in which we are working. The volume of a set in R^2 is usually referred to as its area and the volume of a set in R^1 as its length. We shall later define the area of a set in a different manner, but there should be no confusion between the two quite distinct usages of the word. The properties of Lebesgue measure that we use may be found in any standard text-book in the subject. The following properties are particularly important:

(i) $V_n(\mathfrak{X})>0$ if $\mathfrak{X}$ contains a sphere in R^n. $V_n(\mathfrak{X})=0$ if $\mathfrak{X}$ is contained in an $(n-1)$-dimensional linear manifold of R^n, i.e. the subset of points $\mathbf{x}=(x_1,\ldots,x_n)$ which satisfy an equation of the form

$$a_1x_1+\ldots+a_nx_n=b,$$

where not all the a_i are zero. Such a set is also referred to as a *hyperplane*.

(ii) If $\mathfrak{X}\supset\mathfrak{Y}$ then $V(\mathfrak{X})\geqslant V(\mathfrak{Y})$.

(iii) Let $\mathbf{u}$ be a fixed unit vector and $\mathfrak{P}(\lambda)$ be the hyperplane $\mathbf{x}.\mathbf{u}=\lambda$; then

$$V(\mathfrak{X})=\int_{-\infty}^{+\infty} V_{n-1}(\mathfrak{P}(\lambda)\cap\mathfrak{X})\,d\lambda,$$

where the integral is a Lebesgue integral.

(iv) If $\mathfrak{X}_1$ is similar to $\mathfrak{X}_2$ and the ratio of similitude is $\mu:1$ then

$$V(\mathfrak{X}_1)=\mu^n V(\mathfrak{X}_2).$$

We shall consistently ignore questions of measurability. The sets which are discussed will, in fact, all be measurable but no proofs will be given.

It is assumed that the reader is familiar with the simpler concepts of topology.

The expression $f(\mathbf{x})=a_1\mathbf{x}_1+\ldots+a_n\mathbf{x}_n-b$ where not all the numbers a_i are zero, defines a hyperplane $\mathfrak{P}=\{\mathbf{x}:f(\mathbf{x})=0\}$. The two sets $\{\mathbf{x}:f(\mathbf{x})>0\}$ and $\{\mathbf{x}:f(\mathbf{x})<0\}$ are called open half-spaces and are said to be bounded by $\mathfrak{P}$. Similarly, the sets $\{\mathbf{x}:f(\mathbf{x})\geqslant 0\}$ and $\{\mathbf{x}:f(\mathbf{x})\leqslant 0\}$ are closed half-spaces and are also said to be bounded by $\mathfrak{P}$. If two subsets of R^n, say $\mathfrak{X}_1$ and $\mathfrak{X}_2$, are such that one of the open half-spaces bounded by $\mathfrak{P}$ contains $\mathfrak{X}_1$ and the other open half-space contains $\mathfrak{X}_2$, we say that $\mathfrak{P}$ *separates* the sets $\mathfrak{X}_1$ and $\mathfrak{X}_2$ *strictly*. If one of the closed half-spaces contains $\mathfrak{X}_1$ and the other closed half-space contains $\mathfrak{X}_2$, then we say that $\mathfrak{P}$ *separates* $\mathfrak{X}_1$ and $\mathfrak{X}_2$.

2. The definition of convexity and its relation to affine transformations

DEFINITION. *A set of points $\mathfrak{X}$ is said to be convex if whenever two points $\mathbf{x}_1, \mathbf{x}_2$ belong to $\mathfrak{X}$ all the points of the form*

$$\lambda\mathbf{x}_1+\mu\mathbf{x}_2 \qquad (1\cdot2\cdot1)$$

where $\lambda\geqslant 0$, $\mu\geqslant 0$, $\lambda+\mu=1$ also belong to $\mathfrak{X}$.

Or, to put the definition in a more geometrical language: $\mathfrak{X}$ is convex if whenever it contains two points it also contains the segment joining these two points. This may be put into symbols by writing $\mathfrak{H}(\mathbf{x}_1,\mathbf{x}_2)$ for the closed segment joining the points $\mathbf{x}_1, \mathbf{x}_2$. Then the definition is: $\mathfrak{X}$ is convex if and only if $\mathbf{x}_1\in\mathfrak{X}$, $\mathbf{x}_2\in\mathfrak{X}$ imply $\mathfrak{H}(\mathbf{x}_1,\mathbf{x}_2)\subset\mathfrak{X}$.

Examples of convex sets are: the empty set, a set consisting of a single point, a segment, a sphere, and a regular polygon. To avoid the necessity of considering tiresome trivial but exceptional cases, we make the convention that the empty set is not a convex set, and we shall use the phrase 'convex set' always to mean 'non-empty convex set'.

It should be noted that the definition does not depend upon the dimension of the Euclidean space concerned and that it is possible to use the same definition in much more general spaces. If $\mathfrak{X}_1$ is a convex subset of R^n and $\mathfrak{X}_2$ is a convex subset of R^m then the Cartesian product of $\mathfrak{X}_1, \mathfrak{X}_2$ denoted by $\mathfrak{X}_1 \times \mathfrak{X}_2$ and defined by

$$\mathfrak{X}_1 \times \mathfrak{X}_2 = \{(x_1, \ldots, x_n, x_{n+1}, \ldots, x_{n+m}) : (x_1, \ldots, x_n) \in \mathfrak{X}_1, (x_{n+1}, \ldots, x_{n+m}) \in \mathfrak{X}_2\}$$

is a convex subset of $\mathsf{R}^n \times \mathsf{R}^m$. We identify this last space with R^{n+m}.

The definition can be put in many different forms; some of these will be discussed later. It is often convenient to use a somewhat more general form of the convexity conditions which may be stated as follows. If $\mathfrak{X}$ is a convex set and if $\mathbf{x}_1, \ldots, \mathbf{x}_k$ are k points of $\mathfrak{X}$, then every point $\mathbf{x}$ of the form

$$\mathbf{x} = \lambda_1 \mathbf{x}_1 + \ldots + \lambda_k \mathbf{x}_k \quad (\lambda_i \geqslant 0, \lambda_1 + \ldots + \lambda_k = 1) \qquad (1\cdot2\cdot2)$$

also belongs to $\mathfrak{X}$. For $k = 2$ this is just the definition that $\mathfrak{X}$ is convex. To prove the truth of (1·2·2) assume inductively that it is true for $k = m$ and consider a point $\mathbf{x}$ of the form

$$\mathbf{x} = \lambda_1 \mathbf{x}_1 + \ldots + \lambda_m \mathbf{x}_m + \lambda_{m+1} \mathbf{x}_{m+1} \quad (\lambda_i \geqslant 0, \lambda_1 + \ldots + \lambda_{m+1} = 1). \qquad (1\cdot2\cdot3)$$

If $\lambda_{m+1} = 1$ then $\mathbf{x} = \mathbf{x}_{m+1}$ which belongs to $\mathfrak{X}$ and there is nothing further to prove. Suppose then that $\lambda_{m+1} < 1$. In this case

$$\lambda_1 + \ldots + \lambda_m = 1 - \lambda_{m+1} > 0,$$

and we may put $\mathbf{x}$ in the form

$$\mathbf{x} = (\lambda_1 + \ldots + \lambda_m)\left(\frac{\lambda_1}{\lambda_1 + \ldots + \lambda_m} \mathbf{x}_1 + \ldots + \frac{\lambda_m}{\lambda_1 + \ldots + \lambda_m} \mathbf{x}_m\right) + \lambda_{m+1} \mathbf{x}_{m+1}. \qquad (1\cdot2\cdot4)$$

By the inductive hypothesis the point

$$\mathbf{z} = \frac{\lambda_1}{\lambda_1 + \ldots + \lambda_m}\mathbf{x}_1 + \ldots + \frac{\lambda_m}{\lambda_1 + \ldots + \lambda_m}\mathbf{x}_m$$

belongs to $\mathfrak{X}$. Since $\mathfrak{X}$ is convex and contains both $\mathbf{z}$ and $\mathbf{x}_{m+1}$, it follows from (1·2·4) that it contains $\mathbf{x}$. Thus the inductive hypothesis is proved when $k = m+1$. Hence it is true for all k.

In many ways it is more correct to regard convexity as a property of affine geometry rather than as a property of Euclidean geometry. We do not adopt this point of view here, since there

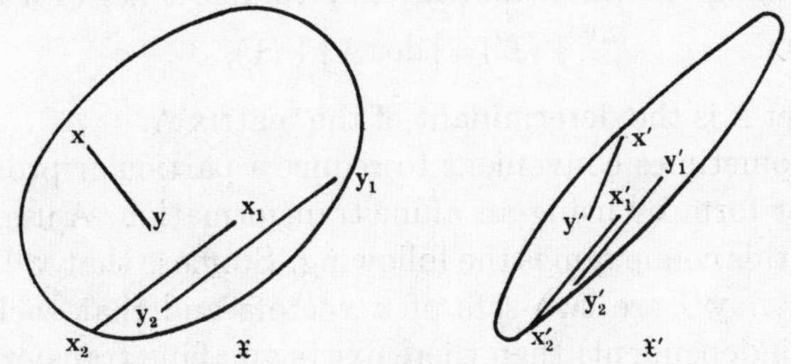

Fig. 1

are some properties of convex sets which are not invariant under affine transformations. But it is important to see what effect an affine transformation will have on the convex subsets of R^n.

A point transformation from $\mathbf{x}$ to $\mathbf{x}'$ is said to be affine if there exists a relation of the form

$$\mathbf{x}' = \mathrm{A}\mathbf{x} + \mathbf{b}, \qquad (1·2·5)$$

where A is a non-singular $n \times n$ real matrix, and $\mathbf{b}$ is a vector of R^n.

Now if $\mathbf{z} = \lambda\mathbf{x} + \mu\mathbf{y}$, where $\lambda \geqslant 0$, $\mu \geqslant 0$, $\lambda + \mu = 1$, and if $\mathbf{x}, \mathbf{y}, \mathbf{z}$ are transformed by (1·2·5) into $\mathbf{x}', \mathbf{y}', \mathbf{z}'$ respectively, then

$$\mathbf{z}' = \lambda\mathbf{x}' + \mu\mathbf{y}'.$$

Thus the segment $\mathfrak{H}(\mathbf{x}, \mathbf{y})$ is transformed into the segment $\mathfrak{H}(\mathbf{x}', \mathbf{y}')$ and a convex set $\mathfrak{X}$ is transformed into another convex set $\mathfrak{X}'$ (see fig. 1).

If
$$\mathbf{x}_1 - \mathbf{x}_2 = \lambda(\mathbf{y}_1 - \mathbf{y}_2),$$
and if $\mathbf{x}_1, \mathbf{x}_2, \mathbf{y}_1, \mathbf{y}_2$ are transformed respectively into $\mathbf{x}_1', \mathbf{x}_2', \mathbf{y}_1', \mathbf{y}_2'$, then
$$\mathbf{x}_1' - \mathbf{x}_2' = \lambda(\mathbf{y}_1' - \mathbf{y}_2').$$

Thus parallel lines are transformed into parallel lines and the ratio of the lengths of segments lying in parallel lines is not altered by the transformation. A similar statement is true of areas lying in parallel planes, etc. The actual change in the length of the segment depends upon the transformation, and in general varies as the inclination of the segment to the axes of coordinates is varied.

The change in the n-dimensional volume $V(\mathfrak{X})$ of a set $\mathfrak{X}$ is given by
$$V(\mathfrak{X}') = |\det \mathrm{A}|\, V(\mathfrak{X}),$$
where $\det \mathrm{A}$ is the determinant of the matrix A.

It is sometimes convenient to reduce a particular problem to a simpler form by using an affine transformation. A useful property in this connexion is the following. Suppose that $\mathbf{x}^{(1)}, \ldots, \mathbf{x}^{(n)}$ and $\mathbf{y}^{(1)}, \ldots, \mathbf{y}^{(n)}$ are two sets of n vectors and that each set is linearly independent; then there exists an affine transformation which sends $\mathbf{x}^{(i)}$ into $\mathbf{y}^{(i)}$, $i = 1, 2, \ldots, n$, and $\mathbf{O}$ into $\mathbf{O}$. To establish the correctness of this statement let L be the $n \times n$ matrix whose columns are $\mathbf{x}^{(1)}, \ldots, \mathbf{x}^{(n)}$ and M be the $n \times n$ matrix whose columns are $\mathbf{y}^{(1)}, \ldots, \mathbf{y}^{(n)}$. Since L is non-singular we can define a matrix
$$\mathrm{A} = \mathrm{M}\mathrm{L}^{-1}.$$
Since M is non-singular, it follows that A is a real non-singular matrix. It is the matrix of the required affine transformation of the form (1·2·5) with $\mathbf{b} = \mathbf{O}$.

If $\mathbf{x}^{(1)}, \ldots, \mathbf{x}^{(n+1)}$ are the vertices of a non-degenerate simplex in R^n and if $\mathbf{y}^{(1)}, \ldots, \mathbf{y}^{(n+1)}$ are those of a regular non-degenerate simplex, then there exists an affine transformation which transforms $\mathbf{x}^{(i)}$ into $\mathbf{y}^{(i)}$, $i = 1, 2, \ldots, n+1$. For the two sets of n vectors $\mathbf{x}^{(i)} - \mathbf{x}^{(1)}$ and $\mathbf{y}^{(i)} - \mathbf{y}^{(1)}$, $i = 2, \ldots, n+1$, are each linearly independent. Thus there exists an affine transformation of the form (1·2·5) which transforms $\mathbf{x}^{(i)} - \mathbf{x}^{(1)}$ into $\mathbf{y}^{(i)} - \mathbf{y}^{(1)}$. Let the matrix of this transformation be A. Then
$$\mathbf{y}^{(i)} - \mathbf{y}^{(1)} = \mathrm{A}(\mathbf{x}^{(i)} - \mathbf{x}^{(1)}) \quad (2 \leqslant i \leqslant n+1). \tag{1·2·6}$$

Define the vector $\mathbf{b}$ by

$$\mathbf{b} = -\mathrm{A}\mathbf{x}^{(1)} + \mathbf{y}^{(1)}. \tag{1·2·7}$$

(1·2·6) and (1·2·7) together imply that

$$\mathbf{y}^{(i)} = \mathrm{A}\mathbf{x}^{(i)} + \mathbf{b} \quad (1 \leqslant i \leqslant n+1).$$

This is the required transformation of the form (1·2·5).

Degenerate affine transformations which arise when the matrix A is singular can also be used. In particular if A is an $m \times n$ matrix with $m < n$, and if A is of rank m then the transformation $\mathbf{x}' = \mathrm{A}.\mathbf{x}$, where $\mathbf{x} \in \mathsf{R}^n$ and $\mathbf{x}' \in \mathsf{R}^m$ defines an affine transformation of R^n onto R^m. This transformation also preserves convexity.

If $\mathfrak{X}_1$ and $\mathfrak{X}_2$ are any two sets in R^n, we can define a third set $\mathfrak{X}_3$, which we denote by $\lambda\mathfrak{X}_1 + \mu\mathfrak{X}_2$, by the relation

$$\mathfrak{X}_3 = \{\mathbf{x} : \mathbf{x} = \lambda\mathbf{x}_1 + \mu\mathbf{x}_2, \mathbf{x}_1 \in \mathfrak{X}_1, \mathbf{x}_2 \in \mathfrak{X}_2\}.$$

The set $\mathfrak{X}_3$ is again convex. This fact may either be proved directly or can be obtained by using a degenerate affine transformation as follows. $\mathfrak{X}_1 \times \mathfrak{X}_2$ is a convex subset of R^{2n} and the $n \times 2n$ matrix A whose elements a_{ij} are defined by $a_{ii} = \lambda$, $a_{i,n+i} = \mu$, $a_{ij} = 0, j \neq i, n+i$, where $1 \leqslant i \leqslant n$, $1 \leqslant j \leqslant 2n$ gives rise to an affine mapping of R^{2n} onto R^n and of $\mathfrak{X}_1 \times \mathfrak{X}_2$ onto $\mathfrak{X}_3$. Thus $\mathfrak{X}_3$ is convex.

Exercises 1·2

1. $\mathfrak{X}$ is a convex set in R^n and $\mathbf{y}$ does not belong to $\mathfrak{X}$. $\mathfrak{H}(\mathbf{y}, \mathfrak{X})$ is the point-set union of all the segments one of whose end-points is $\mathbf{y}$ and the other is a point of $\mathfrak{X}$. Show that $\mathfrak{H}(\mathbf{y}, \mathfrak{X})$ is convex.

2. $\mathfrak{X}_1$ and $\mathfrak{X}_2$ are two disjoint convex sets in R^n and $\mathbf{y}$ does not belong to $\mathfrak{X}_1 \cup \mathfrak{X}_2$. With the notation of Exercise 1·2(1) show that either $\mathfrak{H}(\mathbf{y}, \mathfrak{X}_1) \cap \mathfrak{X}_2 = \phi$ or $\mathfrak{H}(\mathbf{y}, \mathfrak{X}_2) \cap \mathfrak{X}_1 = \phi$.

3. In R^n, n hyperplanes pass through the origin and the origin is their only common point. Show that there is an affine transformation which transforms these hyperplanes into hyperplanes that are perpendicular to one another.

3. Intersections, closures and interiors of convex sets

In dealing with point-sets of a general nature it is usually essential to make some assertion as to whether they are open or closed or neither, and the properties of these three classes of sets are widely different. In the case of convex sets, although a given convex set may be neither open nor closed, yet, since both its closure and its interior (if this is non-empty) are also convex, we can infer the properties of general convex sets from those of closed convex sets or from those of open convex sets. This is usually so simple a matter that it is customary to develop the detailed theory of convex sets for closed sets only. There are some problems which cannot be simplified in this fashion, and for this reason we do not restrict our attention to closed sets until we reach Chapter 4.

In the present section we establish the convexity of the closure of a convex set and the convexity of the non-empty interior of a convex set.

THEOREM 1. *If* $\mathfrak{J}$ *is an index set, if* $\mathfrak{X}_i$ *for each* $i \in \mathfrak{J}$ *is convex and if the intersection* $\bigcap_{i \in \mathfrak{J}} \mathfrak{X}_i$ *is non-empty, then this intersection is convex.*

If $\mathbf{x}$ and $\mathbf{y}$ belong to $\bigcap_{i \in \mathfrak{J}} \mathfrak{X}_i$, they also belong to $\mathfrak{X}_i$. Since $\mathfrak{X}_i$ is convex $\mathfrak{H}(\mathbf{x}, \mathbf{y}) \subset \mathfrak{X}_i$, and since this is true for every i, $i \in \mathfrak{J}$, $\mathfrak{H}(\mathbf{x}, \mathbf{y}) \subset \bigcap_{i \in \mathfrak{J}} \mathfrak{X}_i$. Thus $\bigcap_{i \in \mathfrak{J}} \mathfrak{X}_i$ is convex.

COROLLARY. *If* $(\mathfrak{X}_i)$ *is a sequence of convex sets, then*

$$\liminf \mathfrak{X}_i = \bigcup_{k=1}^{\infty} \bigcap_{i=k}^{\infty} \mathfrak{X}_i$$

is convex or is empty.

For by the theorem each set $\bigcap_{i=k}^{\infty} \mathfrak{X}_i$ is convex or empty. If each of these sets is empty then so is $\liminf \mathfrak{X}_i$. If one of them is not empty then $\liminf \mathfrak{X}_i$ is the union of an increasing sequence of convex sets and as such is easily seen to be convex.

THEOREM 2. *If* $\mathfrak{X}$ *is a convex set and* δ *a positive number, then the set* $\mathfrak{U}(\mathfrak{X}, \delta)$ *is also convex.*

Since $\mathfrak{U}(\mathfrak{X}, \delta)$ contains $\mathfrak{X}$ it is non-empty. Let $\mathbf{y}$ and $\mathbf{y}'$ be two points of $\mathfrak{U}(\mathfrak{X}, \delta)$ (see fig. 2). Then since

$$\mathfrak{U}(\mathfrak{X}, \delta) = \{\mathbf{z} : \rho(\mathbf{z}, \mathfrak{X}) < \delta\},$$

there are at least two points $\mathbf{x}$ and $\mathbf{x}'$ of $\mathfrak{X}$ such that

$$|\mathbf{x}-\mathbf{y}|<\delta, \quad |\mathbf{x}'-\mathbf{y}'|<\delta.$$

Let $\mathbf{y}''$ be a point of $\mathfrak{H}(\mathbf{y},\mathbf{y}')$, say

$$\mathbf{y}''=\lambda\mathbf{y}+\mu\mathbf{y}', \tag{1·3·1}$$

where $\lambda\geqslant 0$, $\mu\geqslant 0$, $\lambda+\mu=1$, and define $\mathbf{x}''$ to be the point

$$\mathbf{x}''=\lambda\mathbf{x}+\mu\mathbf{x}' \tag{1·3·2}$$

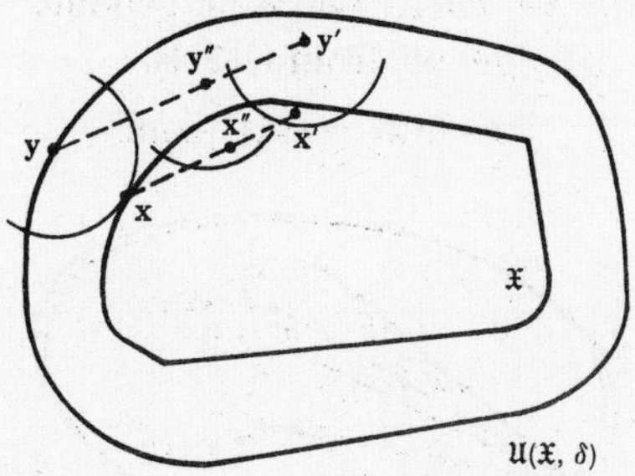

Fig. 2

with the same values of λ and μ as in (1·3·1). Then $\mathbf{x}\in\mathfrak{X}$ (since $\mathfrak{X}$ is convex) and

$$\begin{aligned}|\mathbf{y}''-\mathbf{x}''| &\leqslant |\lambda\mathbf{x}-\lambda\mathbf{y}|+|\mu\mathbf{x}'-\mu\mathbf{y}'| \\ &=\lambda|\mathbf{x}-\mathbf{y}|+\mu|\mathbf{x}'-\mathbf{y}'| \\ &<\delta.\end{aligned}$$

Hence $\mathbf{y}''\in\mathfrak{U}(\mathfrak{X},\delta)$, and it follows that $\mathfrak{U}(\mathfrak{X},\delta)$ is convex.

COROLLARY. *If $\mathfrak{X}$ is convex so is $\overline{\mathfrak{X}}$, the closure of $\mathfrak{X}$.*

For if $\{\delta_i\}$ is a sequence of positive numbers decreasing to zero, then

$$\overline{\mathfrak{X}}=\bigcap_{i=1}^{\infty}\mathfrak{U}(\mathfrak{X},\delta_i),$$

and the result follows from Theorems 1 and 2.

THEOREM 3. *Let $\mathfrak{X}$ be a convex set with a non-void interior $\mathfrak{X}^0$, and let $\mathbf{x}_1$, $\mathbf{x}_2$ be two points of $\mathfrak{X}$, of which $\mathbf{x}_2$ belongs to $\mathfrak{X}^0$. Then every point of $\mathfrak{H}(\mathbf{x}_1,\mathbf{x}_2)$, except possibly $\mathbf{x}_1$, is an interior point of $\mathfrak{X}$.*

We know that every point of $\mathfrak{H}(\mathbf{x}_1, \mathbf{x}_2)$ belongs to $\mathfrak{X}$, since $\mathfrak{X}$ is convex. We have to show that each such point, apart from $\mathbf{x}_1$, is also an interior point of $\mathfrak{X}$.

Since $\mathbf{x}_2$ is an interior point of $\mathfrak{X}$ there is a positive number δ such that

$$\mathfrak{S}(\mathbf{x}_2, \delta) \subset \mathfrak{X}. \tag{1·3·3}$$

Let $\mathbf{y}$ be a point of $\mathfrak{H}(\mathbf{x}_1, \mathbf{x}_2)$, not identical with $\mathbf{x}_1$, say

$$\mathbf{y} = \lambda \mathbf{x}_1 + \mu \mathbf{x}_2, \tag{1·3·4}$$

where $\lambda \geqslant 0$, $\mu > 0$, $\lambda + \mu = 1$. Let $\mathbf{z}$ be a point of $\mathfrak{S}(\mathbf{y}, \mu\delta)$ (see fig. 3). Then $|\mathbf{z} - \mathbf{y}| < \mu\delta$, or, from (1·3·4),

$$|\mathbf{z} - (\lambda \mathbf{x}_1 + \mu \mathbf{x}_2)| < \mu\delta.$$

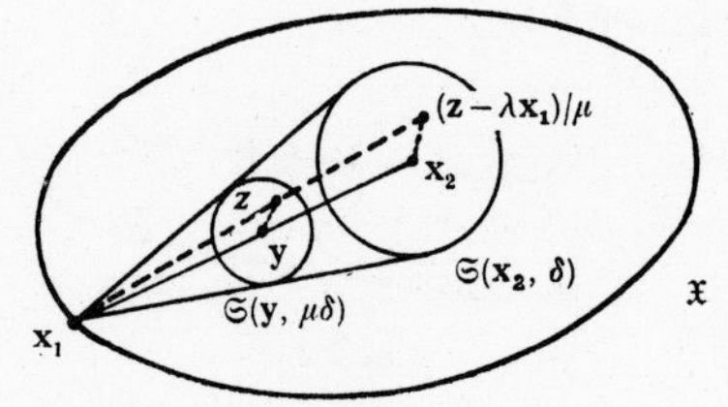

Fig. 3

Since $\mu > 0$, this inequality implies

$$|(\mathbf{z} - \lambda \mathbf{x}_1)/\mu - \mathbf{x}_2| < \delta. \tag{1·3·5}$$

From (1·3·3) and (1·3·5) it follows that $(\mathbf{z} - \lambda \mathbf{x}_1)/\mu$ belongs to $\mathfrak{X}$. But since $\mathfrak{X}$ is convex and since $\mathbf{z}$ can be written in the form

$$\mathbf{z} = \mu[(\mathbf{z} - \lambda \mathbf{x}_1)/\mu] + \lambda \mathbf{x}_1,$$

it follows that $\mathbf{z} \in \mathfrak{X}$.

Thus $\mathfrak{S}(\mathbf{y}, \mu\delta) \subset \mathfrak{X}$, and hence $\mathbf{y}$ is an interior point of $\mathfrak{X}$.

Corollary 1. *If $\mathfrak{X}$ is convex then $\mathfrak{X}^0$, the interior of $\mathfrak{X}$, is either convex or empty.*

Corollary 2. *The theorem is still true if instead of being given that $\mathbf{x}_1 \in \mathfrak{X}$ we are given only that $\mathbf{x}_1 \in \overline{\mathfrak{X}}$.*

Since $\mathbf{x}_2 \in \mathfrak{X}^0$ there exists $\delta > 0$ such that $\mathfrak{S}(\mathbf{x}_2, \delta) \subset \mathfrak{X}^0$. Consider $\mathbf{y} \in \mathfrak{H}(\mathbf{x}_1, \mathbf{x}_2)$ and suppose that $\mathbf{y} \neq \mathbf{x}_1, \mathbf{x}_2$. Let $\mathbf{z}_1$ be a point of $\mathfrak{X}$ that satisfies

$$|\mathbf{z}_1 - \mathbf{x}_1| < \delta |\mathbf{x}_1 - \mathbf{y}| / |\mathbf{x}_2 - \mathbf{y}|.$$

Define $\mathbf{z}_2$ by $\quad \mathbf{z}_2-\mathbf{x}_2=-\dfrac{|\mathbf{x}_2-\mathbf{y}|}{|\mathbf{x}_1-\mathbf{y}|}(\mathbf{z}_1-\mathbf{x}_1);$

then we have

$$\mathbf{y}=\frac{|\mathbf{x}_1-\mathbf{y}|\,\mathbf{x}_2+|\mathbf{x}_2-\mathbf{y}|\,\mathbf{x}_1}{|\mathbf{x}_1-\mathbf{y}|+|\mathbf{x}_2-\mathbf{y}|}=\frac{|\mathbf{x}_1-\mathbf{y}|\,\mathbf{z}_2+|\mathbf{x}_2-\mathbf{y}|\,\mathbf{z}_1}{|\mathbf{x}_1-\mathbf{y}|+|\mathbf{x}_2-\mathbf{y}|}.$$

Thus $\mathbf{y}\in\mathfrak{H}(\mathbf{z}_1,\mathbf{z}_2)$, and we can apply the theorem using $\mathbf{z}_1$, $\mathbf{z}_2$ in place of $\mathbf{x}_1, \mathbf{x}_2$, for by definition $\mathbf{z}_1\in\mathfrak{X}$ and since $|\mathbf{z}_2-\mathbf{x}_2|<\delta$ we have $\mathbf{z}_2\in\mathfrak{X}^0$.

Corollary 3. *If $\mathfrak{X}$ is a convex set and if $\mathfrak{X}^0$ is non-empty, then the closure of $\mathfrak{X}^0$ is identical with $\overline{\mathfrak{X}}$ and the interior of $\overline{\mathfrak{X}}$ is identical with $\mathfrak{X}^0$.*

Since $\mathfrak{X}^0\subset\mathfrak{X}$ we have $\overline{(\mathfrak{X}^0)}\subset\overline{\mathfrak{X}}$. On the other hand, if $\mathbf{x}\in\overline{\mathfrak{X}}$ and $\mathbf{x}_1$ is any point of $\mathfrak{X}^0$, then by the above Corollary 2 the whole of $\mathfrak{H}(\mathbf{x},\mathbf{x}_1)$ is contained in $\mathfrak{X}^0$ except possibly $\mathbf{x}$. Thus $\mathbf{x}\in\overline{(\mathfrak{X}^0)}$ and $\overline{\mathfrak{X}}\subset\overline{(\mathfrak{X}^0)}$.

Next $\overline{\mathfrak{X}}\supset\mathfrak{X}$ and thus $(\overline{\mathfrak{X}})^0\supset\mathfrak{X}^0$. But if $\mathbf{x}\notin\mathfrak{X}^0$ let $\mathbf{x}_1$ be a point of $\mathfrak{X}^0$. On the line $\mathbf{x}_1\mathbf{x}$ any point $\mathbf{y}$ such that $\mathbf{x}\in\mathfrak{H}(\mathbf{x}_1,\mathbf{y})$ is also such that $\mathbf{y}\notin\overline{\mathfrak{X}}$ (or again by Corollary 2 we should have $\mathbf{x}\in\mathfrak{X}^0$). Thus $\mathbf{x}\notin(\overline{\mathfrak{X}})^0$ and $(\overline{\mathfrak{X}})^0\subset\mathfrak{X}^0$.

Remark. Theorem 3 is often used in the following form:

If $\mathfrak{X}$ is convex and $\mathbf{x}_f$ and $\mathbf{x}_i$ are respectively frontier and interior points of $\mathfrak{X}$, then the line through $\mathbf{x}_f$ and $\mathbf{x}_i$ is divided by $\mathbf{x}_f$ into two half-lines of which one does not meet $\mathfrak{X}$ except possibly in the point $\mathbf{x}_f$.

Exercises 1·3

1. $\mathfrak{X}$ is a convex set and δ_i is a positive number for $i=1,2,\ldots,n$. We denote by $T_i(\mathfrak{X})$ the set of points $\mathbf{x}'$ such that there exists a point $\mathbf{x}$ of $\mathfrak{X}$ for which $x'_j=x_j\,(j\neq i)$, $x'_i-x_i\leqslant\delta_i$, where $\mathbf{x}'=(x'_1,\ldots,x'_n)$, $\mathbf{x}=(x_1,\ldots,x_n)$. Show that

(i) $T_i(\mathfrak{X})$ is convex,

(ii) $T_i(\mathfrak{X})$ is closed if $\mathfrak{X}$ is closed,

(iii) $T_i(T_j(\mathfrak{X}))=T_j(T_i(\mathfrak{X}))$.

2. On the two-dimensional surface of a three-dimensional solid sphere a set $\mathfrak{X}$ is said to be convex if it does not contain any diametrically opposite points, and if for any two points $\mathbf{x}_1, \mathbf{x}_2$ of $\mathfrak{X}$ an arc of the great circle joining $\mathbf{x}_1$ to $\mathbf{x}_2$ also belongs to $\mathfrak{X}$. Let $\mathfrak{W}(\mathfrak{X}, \delta)$ be the subset of the sphere formed of those points whose distance from $\mathfrak{X}$ is less than δ. Show that $\mathfrak{W}(\mathfrak{X}, \delta)$ need not be convex even though $\mathfrak{X}$ is convex.

Prove that $\overline{\mathfrak{X}}$ is either convex or contains two diametrically opposite points of the sphere. (An argument similar to that in the Corollary of Theorem 2 no longer applies, but the statement in that corollary is still valid.)

3. Show that a necessary and sufficient condition that a convex set be closed is that its intersection with every straight line should be closed.

4. Sections and projections of convex sets

Many properties of convex sets which are comparatively easy to prove for sets contained in a Euclidean space whose dimension is small can be extended to sets contained in R^n by means of an inductive argument. To do this we need to be able to relate the property concerned in R^n to a similar property in R^{n-1}. There are two methods that may be employed: (*a*) that of forming sections or (*b*) that of projection onto a linear manifold of R^n.

(*a*) If $\mathfrak{P}$ is a hyperplane which intersects the convex subset $\mathfrak{X}$ of R^n, then $\mathfrak{P} \cap \mathfrak{X}$ is called the section of $\mathfrak{X}$ by $\mathfrak{P}$. We regard it as a subset of $\mathfrak{P}$ which is an $n-1$-dimensional linear manifold. Similarly the intersection of $\mathfrak{X}$ by a linear manifold of dimension less than $n-1$ is also referred to as a section of $\mathfrak{X}$ provided it is not empty. By Theorem 1 all sections are convex sets.

(*b*) If $\mathfrak{X}$ is a convex set and $\mathbf{y}$ a point not belonging to $\mathfrak{X}$ every line joining $\mathbf{y}$ to a point $\mathbf{x}$ of $\mathfrak{X}$ is divided by $\mathbf{y}$ into two half-lines, one of which meets $\mathfrak{X}$. The union of all such half-lines is said to be the cone subtended by $\mathfrak{X}$ at $\mathbf{y}$ and is denoted by $\mathfrak{C}(\mathbf{y}, \mathfrak{X})$. It may be verified that $\mathfrak{C}(\mathbf{y}, \mathfrak{X})$ is convex. If $\mathfrak{P}$ is a hyperplane not containing $\mathbf{y}$, the section of $\mathfrak{C}(\mathbf{y}, \mathfrak{X})$ by $\mathfrak{P}$ is called the projection from $\mathbf{y}$ onto $\mathfrak{P}$.

We shall also use projections of convex sets parallel to a fixed direction. If $\mathfrak{X}$ is the given convex set and $\mathbf{l}$ a given unit vector, the set of all lines through points of $\mathfrak{X}$ parallel to $\mathbf{l}$ forms a convex set $\mathfrak{K}(\mathbf{l},\mathfrak{X})$. If $\mathfrak{P}$ is perpendicular to $\mathbf{l}$ the section of $\mathfrak{K}(\mathbf{l},\mathfrak{X})$ by $\mathfrak{P}$ is the projection of $\mathfrak{X}$ in the direction $\mathbf{l}$. It is not defined uniquely, but any property which is congruence invariant is the same for each projection of $\mathfrak{X}$ in the direction $\mathbf{l}$.

The power of these methods will be immediately apparent to anyone familiar with the formidable difficulties associated with topological questions in R^n. These difficulties, which increase so rapidly as the dimension n increases, only affect the theory of convex sets when we come to make a detailed analysis of their structure as in the last two chapters of this book.

EXERCISES 1·4

1. Show that $\mathbf{z} \in \mathfrak{C}(\mathbf{y},\mathfrak{X})$ if and only if $\mathbf{z}$ is of the form

$$\mathbf{z} = \lambda\mathbf{x} + \mu\mathbf{y},$$

where $\lambda \geqslant 0$, $\lambda + \mu = 1$, and deduce that $\mathfrak{C}(\mathbf{y},\mathfrak{X})$ is convex.

2. Show that if $\mathfrak{X}$ is an open convex set and $\mathfrak{Q}$ an r-dimensional linear manifold such that $\mathfrak{Q} \cap \mathfrak{X} \neq \phi$, then the frontier of $\mathfrak{Q} \cap \mathfrak{X}$ relative to $\mathfrak{Q}$ is the intersection of the frontier of $\mathfrak{X}$ with $\mathfrak{Q}$.

3. $\mathfrak{X}$ is a closed convex set of R^n, $n \geqslant 2$. By considering a plane section of $\mathfrak{X}$ passing through the point $\mathbf{y}$, or otherwise, show that if $\mathbf{y} \notin \mathfrak{X}$, then for some $\mathbf{l}$, $\mathbf{y} \notin \mathfrak{K}(\mathbf{l},\mathfrak{X})$. Give an example to show that $\mathbf{y}$ may belong to every set $\mathfrak{K}(\mathbf{l},\mathfrak{X})$ without belonging to $\mathfrak{X}$ if $\mathfrak{X}$ is not closed.

5. The dimension of a convex set. Barycentric coordinates

The interior of a convex set may be empty and we may not be able, immediately, to relate the properties of a convex set to those of its interior. This is because we are regarding the set as situated in an inappropriate Euclidean space, one whose dimension is too large. When we are considering a single convex set there is a considerable advantage in regarding it as a subset of its smallest containing linear manifold. We shall show that if $\mathfrak{L}$ is the smallest linear manifold containing $\mathfrak{X}$, then $\mathfrak{X}$ has interior points relative to $\mathfrak{L}$. This result is proved by using barycentric coordinates.

Of course we cannot make use of this result when considering problems in which more than one convex set is involved. In such a case we cannot usually assume that the convex sets have non-empty interiors. We either have to accept the situation and make such progress as is possible, or we may try to use an approximation argument as explained in Chapter 4.

Let $\mathbf{x}_1, \mathbf{x}_2, \ldots, \mathbf{x}_{n+1}$ be $n+1$ points which form the vertices of a non-degenerate simplex in R^n. The condition for this is that the vectors $\mathbf{x}_2-\mathbf{x}_1$, $\mathbf{x}_3-\mathbf{x}_1, \ldots, \mathbf{x}_{n+1}-\mathbf{x}_1$ should be linearly independent; that is to say, if $\lambda_2, \ldots, \lambda_{n+1}$ are such that

$$\lambda_2(\mathbf{x}_2-\mathbf{x}_1)+\ldots+\lambda_{n+1}(\mathbf{x}_{n+1}-\mathbf{x}_1)=\mathbf{O} \tag{1·5·1}$$

then $\lambda_2=\ldots=\lambda_{n+1}=0$. This condition can also be expressed by saying that if $\lambda_1, \lambda_2, \ldots, \lambda_{n+1}$ are real numbers such that

$$\left.\begin{aligned}\lambda_1\mathbf{x}_1+\lambda_2\mathbf{x}_2+\ldots+\lambda_{n+1}\mathbf{x}_{n+1}&=\mathbf{O},\\ \lambda_1+\lambda_2+\ldots+\lambda_{n+1}&=0,\end{aligned}\right\} \tag{1·5·2}$$

then $$\lambda_1=\lambda_2=\ldots=\lambda_{n+1}=0.$$

Any point $\mathbf{x}$ of R^n can be written in the form

$$\mathbf{x}=\chi_1\mathbf{x}_1+\ldots+\chi_{n+1}\mathbf{x}_{n+1} \quad (1=\chi_1+\ldots+\chi_{n+1}). \tag{1·5·3}$$

The numbers $\chi_1, \ldots, \chi_{n+1}$ are called the barycentric coordinates of $\mathbf{x}$ relative to the basis $\mathbf{x}_1, \ldots, \mathbf{x}_{n+1}$. They are uniquely defined by $\mathbf{x}$ and by the basis $\mathbf{x}_1, \ldots, \mathbf{x}_{n+1}$. To see that these statements are true one has only to observe that the $n+1$ equations of (1·5·3) in $n+1$ unknowns $\chi_1, \ldots, \chi_{n+1}$ have a determinant whose value is the same as that of the $n+1$ equations (1·5·2) in the $n+1$ unknowns $\lambda_1, \lambda_2, \ldots, \lambda_{n+1}$ and is thus non-zero.

Definition. *The linear dimension of a convex set $\mathfrak{X}$ is the largest integer r such that $\mathfrak{X}$ contains $r+1$ points $\mathbf{x}_1, \mathbf{x}_2, \ldots, \mathbf{x}_{r+1}$ for which the vectors $\mathbf{x}_2-\mathbf{x}_1, \ldots, \mathbf{x}_{r+1}-\mathbf{x}_1$ are linearly independent.*

Since this type of dimension is the only one that is used in the sequel we shall drop the adjective 'linear' and refer simply to the 'dimension' of $\mathfrak{X}$.

For such a set $\mathfrak{X}$, every point $\mathbf{x}\in\mathfrak{X}$ is of the form

$$\mathbf{x}=\chi_1\mathbf{x}_1+\ldots+\chi_{r+1}\mathbf{x}_{r+1} \quad (1=\chi_1+\ldots+\chi_{r+1}), \tag{1·5·4}$$

To see this we add $n-r$ points $\mathbf{x}_{r+2}, \ldots, \mathbf{x}_{n+1}$ so that the set $\mathbf{x}_1, \ldots, \mathbf{x}_{n+1}$ forms a basis in R^n. Then $\mathbf{x}$ can be written uniquely in the form

$$\mathbf{x} = \chi_1 \mathbf{x}_1 + \ldots + \chi_{r+1} \mathbf{x}_{r+1} + \chi_{r+2} \mathbf{x}_{r+2} + \ldots + \chi_{n+1} \mathbf{x}_{n+1} \qquad (1 = \chi_1 + \ldots + \chi_{n+1}), \quad (1 \cdot 5 \cdot 5)$$

and (1·5·4) will follow if we can show that

$$\chi_{r+2} = \ldots = \chi_{n+1} = 0$$

Suppose that one of the numbers $\chi_{r+2}, \ldots, \chi_{n+1}$ is non-zero. Since $\mathbf{x}_2 - \mathbf{x}_1, \ldots, \mathbf{x}_{r+1} - \mathbf{x}_1, \mathbf{x} - \mathbf{x}_1$ are linearly dependent, there exist constants $\lambda_2, \ldots, \lambda_{r+1}, \lambda$ not all zero such that

$$\lambda_2(\mathbf{x}_2 - \mathbf{x}_1) + \ldots + \lambda_{r+1}(\mathbf{x}_{r+1} - \mathbf{x}_1) + \lambda(\mathbf{x} - \mathbf{x}_1) = \mathbf{O}. \quad (1 \cdot 5 \cdot 6)$$

Now $\lambda \neq 0$ or the vectors $\mathbf{x}_2 - \mathbf{x}_1, \ldots, \mathbf{x}_{r+1} - \mathbf{x}$, would not be linearly independent. Substitute for $\mathbf{x}$ from (1·5·5) in (1·5·6). We obtain

$$\begin{aligned}&[\lambda(\chi_1 - 1) - \lambda_2 - \ldots - \lambda_{r+1}]\,\mathbf{x}_1 \\ &+ (\lambda\chi_2 + \lambda_2)\,\mathbf{x}_2 + \ldots + (\lambda\chi_{r+1} + \lambda_{r+1})\,\mathbf{x}_{r+1} \\ &+ \lambda\chi_{r+2}\mathbf{x}_{r+2} + \ldots + \lambda\chi_{n+1}\mathbf{x}_{n+1} = \mathbf{O}. \qquad (1 \cdot 5 \cdot 7)\end{aligned}$$

Since

$$[\lambda(\chi_1 - 1) - \lambda_2 - \ldots - \lambda_{r+1}] + (\lambda\chi_2 + \lambda_2) + \ldots + (\lambda\chi_{r+1} + \lambda_{r+1}) + \lambda\chi_{r+2} + \ldots + \lambda\chi_{n+1} = 0$$

and one of $\lambda\chi_{r+2}, \ldots, \lambda\chi_{n+1}$ is non-zero, (1·5·7) contradicts the fact that $\mathbf{x}_1, \mathbf{x}_2, \ldots, \mathbf{x}_{r+1}, \mathbf{x}_{r+2}, \ldots, \mathbf{x}_{n+1}$ form a basis in R^n. Thus each χ_i, $r + 2 \leqslant i \leqslant n + 1$, is zero.

The set of all points of the form (1·5·4) forms a linear manifold which is of dimension r and is denoted by $\mathfrak{L}(\mathfrak{X})$. We define the relative interior and relative frontier of $\mathfrak{X}$ to mean the interior and frontier of $\mathfrak{X}$ relative to $\mathfrak{L}(\mathfrak{X})$. We shall also say that $\mathfrak{L}(\mathfrak{X})$ is the linear manifold carried by $\mathfrak{X}$, or that it is the linear manifold spanned by $\mathfrak{X}$.

$\mathfrak{L}(\mathfrak{X})$ is uniquely defined by $\mathfrak{X}$. For if this were not the case there would be two r-dimensional linear manifolds $\mathfrak{L}_1$ and $\mathfrak{L}_2$, both containing $\mathfrak{X}$ and distinct from one another. Then $\mathfrak{L}_1 \cap \mathfrak{L}_2$ is a linear manifold of dimension less than r which contains $\mathfrak{X}$. Since $\mathfrak{X}$ is of dimension r this is impossible.

THEOREM 4. *The relative interior of a convex set is non-empty.*

For suppose that $\mathbf{x}_1, \ldots, \mathbf{x}_{r+1}$ are points of the convex set $\mathfrak{X}$ such that $\mathbf{x}_1, \ldots, \mathbf{x}_{r+1}$ form a basis of $\mathfrak{L}(\mathfrak{X})$, i.e. that $\mathbf{x} \in \mathfrak{L}(\mathfrak{X})$ if and only if

$$\mathbf{x} = \chi_1 \mathbf{x}_1 + \ldots + \chi_{r+1} \mathbf{x}_{r+1} \quad (1 = \chi_1 + \ldots + \chi_{r+1}). \qquad (1\cdot5\cdot8)$$

Now write

$$\mathbf{x}_0 = (\mathbf{x}_1 + \ldots + \mathbf{x}_{r+1})/(r+1).$$

Then $\mathbf{x}_0 \in \mathfrak{L}(\mathfrak{X})$ and, by (1·2·2), $\mathbf{x}_0 \in \mathfrak{X}$. If the equations (1·5·8) are solved for $\chi_1, \ldots, \chi_{r+1}$, the solutions show that each χ_i depends continuously upon the coordinates of $\mathbf{x}$. Thus there is a positive number δ such that if $\mathbf{x} \in \mathfrak{S}(\mathbf{x}_0, \delta) \cap \mathfrak{L}(\mathfrak{X})$, then each χ_i is positive. Hence, by (1·2·2), each of these points belongs to $\mathfrak{X}$. Thus $\mathbf{x}_0$ is a point of the relative interior of $\mathfrak{X}$.

If $\mathfrak{X}$ is a single point the linear space $\mathfrak{L}(\mathfrak{X})$ and the relative interior of $\mathfrak{X}$ are both the same point.

We have $\mathfrak{L}(\mathfrak{X}_1 \cap \mathfrak{X}_2) \subset \mathfrak{L}(\mathfrak{X}_1) \cap \mathfrak{L}(\mathfrak{X}_2)$. For $\mathfrak{X}_1 \supset \mathfrak{X}_1 \cap \mathfrak{X}_2$ implies that $\mathfrak{L}(\mathfrak{X}_1) \supset \mathfrak{L}(\mathfrak{X}_1 \cap \mathfrak{X}_2)$, and similarly $\mathfrak{L}(\mathfrak{X}_2) \supset \mathfrak{L}(\mathfrak{X}_1 \cap \mathfrak{X}_2)$. In this result the inclusion sign cannot be replaced by an equality unless the relative interior of $\mathfrak{X}_1$ meets the relative interior of $\mathfrak{X}_2$, in which case we have $\mathfrak{L}(\mathfrak{X}_1 \cap \mathfrak{X}_2) = \mathfrak{L}(\mathfrak{X}_1) \cap \mathfrak{L}(\mathfrak{X}_2)$.

Also $\mathfrak{L}(\mathfrak{X}) = \mathfrak{L}(\overline{\mathfrak{X}})$ and further the relative interior of $\mathfrak{X}$ coincides with the relative interior of $\overline{\mathfrak{X}}$. This follows from Theorem 3, Corollary 3, applied to $\mathfrak{X}$ regarded as a subset of $\mathfrak{L}(\mathfrak{X})$.

EXERCISES 1·5

1. Show that if $\mathfrak{X}$ is a given convex set, not a single point, the necessary and sufficient condition that the point $\mathbf{y}$ should belong to $\mathfrak{L}(\mathfrak{X})$ is that some line through $\mathbf{y}$ should meet $\mathfrak{X}$ in a non-degenerate segment.

2. If $\mathfrak{X}$ is a convex set show that the closure of the relative interior of $\mathfrak{X}$ coincides with the closure of $\mathfrak{X}$.

3. If $\mathfrak{X}$ is a convex set of dimension r, show that the projection of $\mathfrak{X}$ onto a hyperplane from a point, or parallel to a line, is of dimension r or $r-1$.

4. If $\mathbf{x}$ is a relative interior point of a set $\mathfrak{X}$ and $\mathfrak{X}$ is projected from a point $\mathbf{y}$ onto a set $\mathfrak{X}'$, then $\mathbf{x}$ is projected onto a relative interior point of $\mathfrak{X}'$.

6. Intersections of convex sets with hyperplanes

It will be shown later that any n-dimensional bounded closed convex set is homeomorphic to an n-cell or to a solid n-dimensional closed sphere. In this section we obtain a certain amount of information about the topological nature of a convex set. The theorems proved here may be regarded as a counterpart to those in the following section. Here we consider hyperplanes that intersect convex sets, whereas in § 7 we consider hyperplanes that either touch or do not intersect convex sets.

DEFINITION. *A hyperplane* $\mathfrak{P}$ *is said to cut the convex set* $\mathfrak{X}$ *if there are points of* $\mathfrak{X}$ *in each of the two open half-spaces into which* $\mathfrak{P}$ *separates* R^n.

THEOREM 5. $\mathfrak{P}$ *cuts* $\mathfrak{X}$ *if and only if the following two conditions hold:*

(i) $\mathfrak{P} \not\supset \mathfrak{L}(\mathfrak{X})$,

(ii) $\mathfrak{P}$ *intersects the relative interior of* $\mathfrak{X}$.

Suppose that $\mathfrak{P}$ cuts $\mathfrak{X}$. Then $\mathfrak{X}$ is non-void and there are points of $\mathfrak{X}$, say $\mathbf{x}_1, \mathbf{x}_2$, one each in the two open half-spaces bounded by $\mathfrak{P}$. Since $\mathfrak{X} \subset \mathfrak{L}(\mathfrak{X})$ and $\mathbf{x}_1 \notin \mathfrak{P}$ we have $\mathfrak{L}(\mathfrak{X}) \not\subset \mathfrak{P}$ and (i) is true. Next let $\mathbf{x}_i$ be a point of the relative interior of $\mathfrak{X}$. By Theorem 3 applied to $\mathfrak{X}$ regarded as a subset of the space $\mathfrak{L}(\mathfrak{X})$, every point of $\mathfrak{H}(\mathbf{x}_1, \mathbf{x}_i)$ and of $\mathfrak{H}(\mathbf{x}_2, \mathbf{x}_i)$, except possibly $\mathbf{x}_1$ and $\mathbf{x}_2$, belongs to the relative interior of $\mathfrak{X}$. But the broken line $\mathfrak{H}(\mathbf{x}_1, \mathbf{x}_i) \cup \mathfrak{H}(\mathbf{x}_2, \mathbf{x}_i)$ joins points from the two open half-spaces bounded by $\mathfrak{P}$ and must therefore meet $\mathfrak{P}$ in, say, $\mathbf{y}$. The point $\mathbf{y}$ is neither $\mathbf{x}_1$ nor $\mathbf{x}_2$ and thus belongs to the relative interior of $\mathfrak{X}$ and (ii) is proved.

On the other hand, if (i) and (ii) are satisfied, let $\mathbf{y}$ be a point of $\mathfrak{L}(\mathfrak{X})$ not in $\mathfrak{P}$ and $\mathbf{x}_i$ be a point of the relative interior of $\mathfrak{X}$ that lies on $\mathfrak{P}$. Any line through $\mathbf{y}$ meets $\mathfrak{P}$ in at most one point, since otherwise it would lie completely in $\mathfrak{P}$. Thus the line $\mathbf{y}\mathbf{x}_i$ meets $\mathfrak{P}$ only in the point $\mathbf{x}_i$. Since $\mathbf{y}\mathbf{x}_i \subset \mathfrak{L}(\mathfrak{X})$ and $\mathbf{x}_i$ is a point of the relative interior of $\mathfrak{X}$, there is a segment $\mathfrak{H}(\mathbf{x}_1, \mathbf{x}_2)$ of the line $\mathbf{y}\mathbf{x}_i$ which is contained in $\mathfrak{X}$ and is such that $\mathbf{x}_i$ lies between $\mathbf{x}_1$ and $\mathbf{x}_2$. The points $\mathbf{x}_1$ and $\mathbf{x}_2$ lie in different open half-spaces bounded by $\mathfrak{P}$. Thus $\mathfrak{P}$ cuts $\mathfrak{X}$ and the proof of the theorem is complete.

COROLLARY. *If a hyperplane cuts* $\mathfrak{X}$ *it also cuts the relative interior of* $\mathfrak{X}$.

THEOREM 6. *If the hyperplane* $\mathfrak{P}$ *cuts the convex set* $\mathfrak{X}$ *then*

(i) *the intersection of* $\mathfrak{P}$ *with the relative frontier of* $\mathfrak{X}$ *is nowhere dense in the relative frontier of* $\mathfrak{X}$,

(ii) *the intersection of* $\mathfrak{P}$ *with the frontier of* $\mathfrak{X}$ *is nowhere dense in the frontier of* $\mathfrak{X}$.

(i) We have to show that if $\mathbf{x}_f$ is any point of the relative frontier of $\mathfrak{X}$, then in any spherical neighbourhood $\mathfrak{U}(\mathbf{x}_f, \delta)$ of $\mathbf{x}_f$ there is a point of the relative frontier of $\mathfrak{X}$ that does not belong

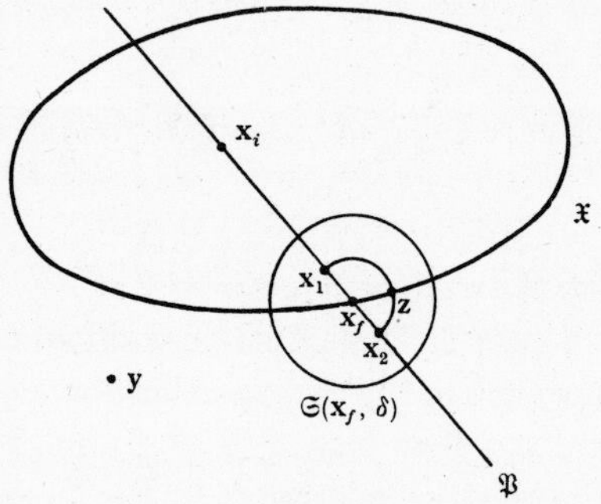

Fig. 4

to $\mathfrak{P}$. If $\mathbf{x}_f \notin \mathfrak{P}$ this is trivially true. Suppose then that $\mathbf{x}_f \in \mathfrak{P}$. Let $\mathbf{x}_i$ be a point of the relative interior of $\mathfrak{X}$ that lies on $\mathfrak{P}$. On the line joining $\mathbf{x}_i$ to $\mathbf{x}_f$ let $\mathbf{x}_1$ and $\mathbf{x}_2$ be two points of $\mathfrak{U}(\mathbf{x}_f, \delta)$ such that $\mathbf{x}_1$ lies between $\mathbf{x}_i$ and $\mathbf{x}_f$, $\mathbf{x}_f$ lies between $\mathbf{x}_1$ and $\mathbf{x}_2$ (see fig. 4). Since $\mathfrak{P}$ cuts $\mathfrak{X}$ there is a point $\mathbf{y}$ of $\mathfrak{L}(\mathfrak{X})$ not lying on $\mathfrak{P}$. Since the line $\mathbf{x}_i\mathbf{x}_f \subset \mathfrak{P}$, $\mathbf{y}$ does not lie on $\mathbf{x}_i\mathbf{x}_f$. Denote the plane in which $\mathbf{x}_i$, $\mathbf{x}_f$ and $\mathbf{y}$ lie by $\mathfrak{B}$. In $\mathfrak{B}$ construct a semicircle with end-points $\mathbf{x}_1, \mathbf{x}_2$. By theorem 3, $\mathbf{x}_1$ is a point of the relative interior of $\mathfrak{X}$ and, by Corollary 2 to that theorem, $\mathbf{x}_2$ is not a point of $\overline{\mathfrak{X}}$. The semicircle joining $\mathbf{x}_1$ to $\mathbf{x}_2$ meets the relative frontier of $\mathfrak{X}$ in at least one point $\mathbf{z}$ which is neither $\mathbf{x}_1$ nor $\mathbf{x}_2$. Then $\mathbf{z} \notin \mathfrak{P}$ and $\mathfrak{U}(\mathbf{x}_f, \delta)$ contains a point of the relative frontier of $\mathfrak{X}$ that does not belong to $\mathfrak{P}$.

This completes the proof of (i).

(ii) If $\mathfrak{X}$ is n-dimensional then (ii) is implied by (i). If $\mathfrak{X}$ is not n-dimensional then $\overline{\mathfrak{X}}$ coincides with $\mathrm{Fr}\,\mathfrak{X}$ and we have to prove that if $\mathbf{x}_1 \in \overline{\mathfrak{X}} \cap \mathfrak{P}$ and δ is any positive number, there is a point $\mathbf{x}_2$ of $\overline{\mathfrak{X}}$ in $\mathfrak{U}(\mathbf{x}_1, \delta)$ that does not lie on $\mathfrak{P}$. Let $\mathbf{x}$ be a point of $\mathfrak{X}$ not on $\mathfrak{P}$. The line $\mathbf{x}\mathbf{x}_1$ meets $\mathfrak{P}$ in the single point $\mathbf{x}_1$. On this line take $\mathbf{x}_2$, $\mathbf{x}_2 \neq \mathbf{x}_1$, such that $\mathbf{x}_2$ lies both between $\mathbf{x}$ and $\mathbf{x}_1$ and inside $\mathfrak{U}(\mathbf{x}_1, \delta)$. Then $\mathbf{x}_2$ belongs to $\overline{\mathfrak{X}}$, does not lie on $\mathfrak{P}$ and lies inside $\mathfrak{U}(\mathbf{x}_1, \delta)$.

Thus (ii) is proved and the theorem is established.

EXERCISES 1·6

1. (*a*) If $\mathfrak{X}$ is a given convex set in R^n and $\mathfrak{X}'$ is its projection onto the hyperplane $\mathfrak{P}$ by lines parallel to $\mathbf{l}$, show that an $(n-2)$-dimensional linear manifold in $\mathfrak{P}$, say $\mathfrak{P}_1$, cuts $\mathfrak{X}'$ if and only if the $(n-1)$-dimensional linear manifold spanned by $\mathfrak{P}_1$ and a vector through a point of $\mathfrak{P}_1$ in the direction $\mathbf{l}$ cuts $\mathfrak{X}$.

(*b*) Let $\mathfrak{X}$ be a given convex set in R^n and $\mathfrak{X}'$ its projection from a point $\mathbf{y}$ onto a hyperplane $\mathfrak{P}$. Show that the $(n-2)$-dimensional linear manifold $\mathfrak{P}_1$ contained in $\mathfrak{P}$ cuts $\mathfrak{X}'$ if and only if the $(n-1)$-dimensional linear manifold through $\mathfrak{P}_1$ and $\mathbf{y}$ cuts $\mathfrak{X}$.

7. Separation of convex sets and support hyperplanes

The theorems proved in this section are of fundamental importance in the theory. The main result is that through every point of the frontier of a convex set there passes a hyperplane which does not cut the convex set. Such a hyperplane is called a support hyperplane. It is also shown that we can take this property as a definition of convexity. We shall refer to this definition as the 'dual' of that given originally in § 2. The concept of duality which occurs here for the first time is not the same as that used in projective geometry or in the theory of Banach spaces. The two dual definitions of convexity lead to 'dual' theorems and to dual proofs of the same theorem. But the duality is not exact and there are severe limitations upon its applicability.

THEOREM 7. *If $\mathfrak{X}$ is an open convex set and $\mathfrak{L}$ is a linear manifold of dimension r, $0 \leqslant r < n$, such that the intersection $\mathfrak{L} \cap \mathfrak{X}$ is empty, then there exists a hyperplane $\mathfrak{P}$ such that $\mathfrak{P} \supset \mathfrak{L}$ and $\mathfrak{P} \cap \mathfrak{X}$ is empty.*

Let $\mathfrak{P}$ be a linear manifold of maximal dimension that contains $\mathfrak{L}$ and does not meet $\mathfrak{X}$. Suppose that $\mathfrak{P}$ is of dimension s. Project R^n onto the linear subspace $\mathfrak{Q}$ that is perpendicular to $\mathfrak{P}$ and whose dimension is $n-s$. $\mathfrak{P}$ projects into a point $\mathbf{p}$ and $\mathfrak{X}$ into an open convex subset $\mathfrak{X}_1$ of $\mathfrak{Q}$. The point $\mathbf{p}$ does not belong to $\mathfrak{X}_1$ and every line through $\mathbf{p}$ in $\mathfrak{Q}$ meets $\mathfrak{X}_1$ by the maximal dimensional property of $\mathfrak{P}$. If $\mathfrak{Q}$ is one-dimensional the theorem is true; suppose then that $\mathfrak{Q}$ is of dimension greater than or equal to 2. Consider a plane through $\mathbf{p}$ in $\mathfrak{Q}$. It meets $\mathfrak{X}_1$ in a convex two-dimensional relatively open set, say $\mathfrak{X}_2$. The half-lines through $\mathbf{p}$ that meet $\mathfrak{X}_2$ and terminate at $\mathbf{p}$ form an open sector. Since $\mathbf{p}\notin\mathfrak{X}_2$ and $\mathfrak{X}_2$ is convex the angle of this sector is $\leqslant\pi$. Since every line through $\mathbf{p}$ meets $\mathfrak{X}_2$ and $\mathfrak{X}_2$ is open the angle of this sector is $>\pi$. This contradiction establishes Theorem 7, for it shows that in fact $\mathfrak{Q}$ is one-dimensional and thus $\mathfrak{P}$ is $(n-1)$-dimensional.

COROLLARY 1. *Let $\mathfrak{X}_1$ be an open convex set and $\mathfrak{X}_2$ a convex set not meeting $\mathfrak{X}_1$, then there exists a hyperplane $\mathfrak{P}$ such that $\mathfrak{P}$ separates $\mathfrak{X}_1$ from $\mathfrak{X}_2$.*

The set $\mathfrak{X}_1-\mathfrak{X}_2$ is open and convex (see the end of § 2). Since $\mathbf{O}$ does not belong to this set there is a hyperplane $\mathfrak{P}$ passing through $\mathbf{O}$ and not meeting $\mathfrak{X}_1-\mathfrak{X}_2$. Suppose that it is given by the equation $\mathbf{a}.\mathbf{x}=0$ and that the signs of the coefficients $a_1, \ldots, a_n$ are such that for $\mathbf{x}\in\mathfrak{X}_1-\mathfrak{X}_2$, $\mathbf{a}.\mathbf{x}>0$. Then for $\mathbf{x}_1\in\mathfrak{X}_1$ and $\mathbf{x}_2\in\mathfrak{X}_2$ we have $\mathbf{a}.\mathbf{x}_1>\mathbf{a}.\mathbf{x}_2$ and the number $\inf\{\mathbf{a}.\mathbf{x}:\mathbf{x}\in\mathfrak{X}_1\}$ exists finitely. Denote it by λ. Then $\mathfrak{X}_1$ and $\mathfrak{X}_2$ are separated by the hyperplane $\mathbf{a}.\mathbf{x}=\lambda$.

DEFINITION. *A hyperplane that intersects the closure of a convex set $\mathfrak{X}$ and does not cut $\mathfrak{X}$ is said to be a support hyperplane of $\mathfrak{X}$.*

We shall say that such a hyperplane supports $\mathfrak{X}$ at the point or points at which it meets $\overline{\mathfrak{X}}$.

THEOREM 8. *Through every point on the frontier of a convex set $\mathfrak{X}$ there passes at least one support hyperplane of $\mathfrak{X}$.*

If $\mathfrak{X}$ is not n-dimensional any hyperplane that contains $\mathfrak{L}(\mathfrak{X})$ is a support hyperplane of $\mathfrak{X}$ at all points of $\mathfrak{X}$.

If $\mathfrak{X}$ is n-dimensional, take $\mathfrak{X}^0$ as the set $\mathfrak{X}_1$ in Corollary 1 to Theorem 7 and the point $\mathbf{p}$ on the frontier of $\mathfrak{X}$ as the set $\mathfrak{X}_2$. The hyperplane $\mathfrak{P}$ that separates $\mathbf{p}$ from $\mathfrak{X}^0$ does not cut $\mathfrak{X}^0$ and therefore does not cut $\mathfrak{X}$; but since $\mathbf{p} \in \overline{(\mathfrak{X}^0)}$ by Corollary 3 to Theorem 3, it follows that $\mathfrak{P}$ contains $\mathbf{p}$. Thus $\mathfrak{P}$ is a support hyperplane of $\mathfrak{X}$ at $\mathbf{p}$.

THEOREM 9. *If the closed set $\mathfrak{X}$ has a non-void interior and if through every point of its frontier there passes a support hyperplane to $\mathfrak{X}$, then $\mathfrak{X}$ is convex.*

For if $\mathfrak{X}$ is not convex then there are two points $\mathbf{x}_1$ and $\mathbf{x}_2$ of $\mathfrak{X}$ such that for some point $\mathbf{y}$ we have

$$\mathbf{y} \in \mathfrak{H}(\mathbf{x}_1, \mathbf{x}_2), \quad \mathbf{y} \notin \mathfrak{X}.$$

Let $\mathbf{z}$ be a point of the interior of $\mathfrak{X}$ that does not lie on the line $\mathbf{x}_1\mathbf{x}_2$. Then there is a point $\mathbf{x}_f$ of the frontier of $\mathfrak{X}$ on $\mathfrak{H}(\mathbf{z}, \mathbf{y})$ and $\mathbf{x}_f$ is an interior point of the triangle $\mathbf{x}_1\mathbf{x}_2\mathbf{z}$. But any hyperplane through $\mathbf{x}_f$ separates one of $\mathbf{x}_1, \mathbf{x}_2, \mathbf{z}$ from at least one of the other two vertices $\mathbf{x}_1, \mathbf{x}_2, \mathbf{z}$. Thus there is no support hyperplane to $\mathfrak{X}$ at $\mathbf{x}_f$. This contradiction proves the theorem.

EXERCISES 1·7

1. Prove that if $\mathfrak{X}_1$ and $\mathfrak{X}_2$ are disjoint closed bounded convex sets there is a hyperplane $\mathfrak{P}$ which separates $\mathfrak{X}_1$ strictly from $\mathfrak{X}_2$. Give an example to show that this result may be untrue if both sets $\mathfrak{X}_1$ and $\mathfrak{X}_2$ are unbounded.

8. The convex cover

With any given point set $\mathfrak{X}$ there can be associated a convex set $\mathfrak{H}(\mathfrak{X})$. This set, the *convex cover* of $\mathfrak{X}$, is the smallest convex set that contains $\mathfrak{X}$. In certain geometrical problems it is possible to replace $\mathfrak{X}$ by $\mathfrak{H}(\mathfrak{X})$, and in all cases $\mathfrak{H}(\mathfrak{X})$ has certain properties which reflect more or less accurately the corresponding properties of $\mathfrak{X}$. For this reason we study the relationship between $\mathfrak{X}$ and $\mathfrak{H}(\mathfrak{X})$. We are led to consider certain points of a convex set, its *extreme points*. These points determine completely the structure of a closed set.

DEFINITION. *If $\mathfrak{X}$ is any set of points in R^n, then by the convex cover of $\mathfrak{X}$ is meant the set of points which is the intersection of all the convex sets that contain $\mathfrak{X}$.*

This set is denoted by $\mathfrak{H}(\mathfrak{X})$. It is convex, and a necessary and sufficient condition that $\mathfrak{X}$ be convex is that $\mathfrak{X} = \mathfrak{H}(\mathfrak{X})$. If $\mathfrak{X}_1, \ldots, \mathfrak{X}_k$ is a finite collection of sets then we denote the convex cover of their union by $\mathfrak{H}(\mathfrak{X}_1, \ldots, \mathfrak{X}_k)$. This notation is consistent with the use of $\mathfrak{H}(\mathbf{x}_1, \mathbf{x}_2)$ for the segment joining $\mathbf{x}_1$ to $\mathbf{x}_2$. If $\mathfrak{Y}$ is the projection of $\mathfrak{X}$ on a linear subspace then $\mathfrak{H}(\mathfrak{Y})$ is the projection of $\mathfrak{H}(\mathfrak{X})$. This follows immediately from the definition.

Theorem 10. *If $\mathfrak{X}$ is closed and bounded then $\mathfrak{H}(\mathfrak{X})$ is closed and bounded.*

That $\mathfrak{H}(\mathfrak{X})$ is bounded follows from the fact that for some positive number R, $\mathfrak{S}(\mathbf{O}, R) \supset \mathfrak{X}$, and since $\mathfrak{S}(\mathbf{O}, R)$ is convex it follows that $\mathfrak{S}(\mathbf{O}, R) \supset \mathfrak{H}(\mathfrak{X})$; thus $\mathfrak{H}(\mathfrak{X})$ is bounded.

The argument used to show that $\mathfrak{H}(\mathfrak{X})$ is closed is similar to that used in Theorem 7. We have to show that the frontier of $\mathfrak{H}(\mathfrak{X})$ is contained in $\mathfrak{H}(\mathfrak{X})$. Suppose that this is not the case and that there is a point $\mathbf{y}$ which belongs to $\overline{\mathfrak{H}(\mathfrak{X})}$ and not to $\mathfrak{H}(\mathfrak{X})$. Of all linear manifolds which contain $\mathbf{y}$ and do not meet $\mathfrak{H}(\mathfrak{X})$, let $\mathfrak{Q}$ be one of maximal dimension. $\mathfrak{Q}$ cannot be $(n-1)$-dimensional, for if it were it would be at a positive distance from the closed bounded set $\mathfrak{X}$, and thus a suitably chosen parallel hyperplane would separate $\mathbf{y}$ strictly from $\mathfrak{X}$. This is impossible, as it implies that $\mathbf{y} \notin \overline{\mathfrak{H}(\mathfrak{X})}$. Project R^n onto the linear space orthogonal to $\mathfrak{Q}$, say $\mathfrak{T}$. Then $\mathfrak{X}$ projects onto a closed set $\mathfrak{Y}$ and $\mathfrak{H}(\mathfrak{X})$ onto $\mathfrak{H}(\mathfrak{Y})$, $\mathfrak{Q}$ projects into a point $\mathbf{q}$, and every line through $\mathbf{q}$ in $\mathfrak{T}$ meets $\mathfrak{H}(\mathfrak{Y})$. Since $\mathfrak{T}$ is of dimension at least 2 we can define $\mathfrak{R}$ to be a plane section of $\mathfrak{T}$ passing through $\mathbf{q}$. Now $\mathfrak{Y}$ is closed and since $\mathbf{q} \notin \mathfrak{Y}$, there exists a positive number δ such that $\mathfrak{S}(\mathbf{q}, \delta) \cap \mathfrak{Y} = \phi$. Form the set of all segments joining $\mathbf{q}$ to points of $\mathfrak{Y}$, say $\mathfrak{K}$. Since $\mathbf{q}$ is a frontier point of $\mathfrak{H}(\mathfrak{Y})$ there is a support line, say l, to $\mathfrak{Y} \cap \mathfrak{R} \cap$ at $\mathbf{q}$ in $\mathfrak{R}$. Thus $\mathfrak{K}$ Fr $(\mathfrak{S}(\mathbf{q}, \delta))$ is a closed set that lies in a closed semicircle. If the least closed arc of Fr $\mathfrak{S}(\mathbf{q}, \delta)$ containing this intersection is a semicircle, then $\mathbf{q} \in \mathfrak{H}(\mathfrak{Y})$. This is impossible. If the least closed arc of Fr $\mathfrak{S}(\mathbf{q}, \delta)$ containing this intersection is less than a semicircle, and its end-points are $\mathbf{a}$, $\mathbf{b}$, then the line through $\mathbf{q}$ parallel to $\mathbf{ab}$ does not meet $\mathfrak{H}(\mathfrak{Y}) \cap \mathfrak{R}$, i.e. does not meet $\mathfrak{H}(\mathfrak{Y})$. This again is impossible by the maximal property of $\mathfrak{Q}$. Thus in either case we are led to a contradiction and the theorem is proved.

Remarks. (i) A shorter proof will be given in Chapter 2 depending on Carathéodory's theorem (see Remark (ii) after Theorem 18).

(ii) We cannot dispense with the condition that $\mathfrak{X}$ be bounded. For example, take $\mathfrak{X}$ to be the two sequences of points in a plane with coordinates

$$(1,1),(2,\tfrac{1}{2}),\ldots,\left(n,\frac{1}{n}\right),\ldots;\quad(-1,1),(-2,\tfrac{1}{2}),\ldots,\left(-n,\frac{1}{n}\right),\ldots.$$

$\mathfrak{X}$ is closed, but $\mathfrak{H}(\mathfrak{X})$ is not closed since it includes all the points $(0,y)$, $0<y\leqslant 1$, but not the point $(0,0)$.

THEOREM 11. *If $\mathfrak{X}$ is bounded the set $\mathfrak{H}(\overline{\mathfrak{X}})$ is the intersection of all the closed half-spaces that contain $\mathfrak{X}$.*

Denote the intersection of all the closed half-spaces that contain $\mathfrak{X}$ by $\mathfrak{H}^*(\mathfrak{X})$. Since $\mathfrak{H}^*(\mathfrak{X})$ is convex and contains $\overline{\mathfrak{X}}$, we have $\mathfrak{H}^*(\mathfrak{X})\supset\mathfrak{H}(\overline{\mathfrak{X}})$.

On the other hand, by Theorems 7 and 10, if $\mathbf{y}$ is a point of the complement of $\mathfrak{H}(\overline{\mathfrak{X}})$ there is a hyperplane $\mathfrak{P}$ that separates $\mathbf{y}$ strictly from $\mathfrak{H}(\overline{\mathfrak{X}})$ (cf. Ex. 1·7). Hence $\mathbf{y}$ does not belong to a certain closed half-space which contains $\mathfrak{H}(\overline{\mathfrak{X}})$. Since this half-space contains $\mathfrak{X}$ it follows that $\mathbf{y}\notin\mathfrak{H}^*(\mathfrak{X})$. Thus $\mathfrak{H}^*(\mathfrak{X})\subset\mathfrak{H}(\overline{\mathfrak{X}})$ and the theorem is proved.

Remarks. (i) The condition that $\mathfrak{X}$ be bounded cannot be removed.

(ii) We can use this theorem to characterize convexity. Let $\mathfrak{X}$ be a closed bounded set such that if $\mathbf{y}$ is any point of the complement of $\mathfrak{X}$ in R^n, then there exists a hyperplane $\mathfrak{P}$ that strictly separates $\mathfrak{X}$ from $\mathbf{y}$. Then $\mathfrak{X}$ is convex.

DEFINITION. *The diameter of a set is the upper bound of the distances between any two points of the set. The diameter of $\mathfrak{X}$ will be denoted by $D(\mathfrak{X})$.*

THEOREM 12. $D(\mathfrak{H}(\mathfrak{X}))=D(\mathfrak{X})$.

Since $\mathfrak{H}(\mathfrak{X})\supset\mathfrak{X}$ we have $D(\mathfrak{H}(\mathfrak{X}))\geqslant D(\mathfrak{X})$. Also if $d>D(\mathfrak{X})$ and $\mathbf{x}_1,\mathbf{x}_2\in\mathfrak{X}$, then $|\mathbf{x}_1-\mathbf{x}_2|<d$, thus $\mathfrak{X}\subset\mathfrak{S}(\mathbf{x}_2,d)$. But $\mathfrak{S}(\mathbf{x}_2,d)$ is convex and thus $\mathfrak{H}(\mathfrak{X})\subset\mathfrak{S}(\mathbf{x}_2,d)$. Let $\mathbf{y}_1,\mathbf{y}_2$ be two points of $\mathfrak{H}(\mathfrak{X})$. Then $\mathbf{y}_1\in\mathfrak{S}(\mathbf{x}_2,d)$ for all $\mathbf{x}_2\in\mathfrak{X}$. Hence $\mathbf{x}_2\in\mathfrak{S}(\mathbf{y}_1,d)$,

i.e. $\mathfrak{X} \subset \mathfrak{S}(\mathbf{y}_1, d)$, and thus, as before, $\mathfrak{H}(\mathfrak{X}) \subset \mathfrak{S}(\mathbf{y}_1, d)$. But this implies that $|\mathbf{y}_1 - \mathbf{y}_2| < d$ and $D(\mathfrak{H}(\mathfrak{X})) \leqslant D(\mathfrak{X})$.

Finally, $D(\mathfrak{H}(\mathfrak{X})) = D(\mathfrak{X})$ and the theorem is proved.

DEFINITION. (i) *A point* $\mathbf{x}$ *is said to be an extreme point of the convex set* $\mathfrak{X}$ *if* $\mathbf{x} \in \mathfrak{X}$ *and there are no two points* $\mathbf{x}_1, \mathbf{x}_2$ *of* $\mathfrak{X}$ *such that*

$$\mathbf{x} \in \mathfrak{H}(\mathbf{x}_1, \mathbf{x}_2) \quad (\mathbf{x} \neq \mathbf{x}_1, \mathbf{x} \neq \mathbf{x}_2).$$

(ii) *A point* $\mathbf{x}$ *is said to be an exposed point of the convex set* $\mathfrak{X}$ *if* $\mathbf{x} \in \mathfrak{X}$ *and there is a support hyperplane to* $\mathfrak{X}$ *that meets* $\mathfrak{X}$ *in the single point* $\mathbf{x}$.

THEOREM 13. (i) *A support hyperplane to a closed bounded convex set* $\mathfrak{X}$ *contains at least one extreme point of* $\mathfrak{X}$.

(ii) *A closed bounded convex set is the closure of the convex cover of its extreme points.*

(i) Let $\mathfrak{P}$ be a support hyperplane to $\mathfrak{X}$ and $\mathfrak{Y}$ the set of extreme points of $\mathfrak{X}$. It may be verified that the extreme points of $\mathfrak{P} \cap \mathfrak{X}$ is the set $\mathfrak{P} \cap \mathfrak{Y}$. Assume inductively that (i) is true for sets of dimension $\leqslant n-1$; then it follows that it is true for sets of dimension n. It is trivially true for sets of dimension 1 and therefore true generally.

(ii) We use the notation of (i); since $\mathfrak{X} \supset \mathfrak{Y}$ and $\mathfrak{X}$ is convex, $\mathfrak{X} \supset \mathfrak{H}(\mathfrak{Y})$. On the other hand, if $\mathbf{x} \in \mathfrak{X}$ and $\mathbf{x} \notin \overline{\mathfrak{H}(\mathfrak{Y})}$, then there is a hyperplane $\mathfrak{P}$ that separates $\mathbf{x}$ strictly from $\mathfrak{H}(\mathfrak{Y})$. The parallel support hyperplane of $\mathfrak{X}$ that is strictly separated from $\mathfrak{H}(\mathfrak{Y})$ by $\mathfrak{P}$ contains a point of $\mathfrak{Y}$ (by (i)). This is impossible and $\mathfrak{X} \subset \overline{\mathfrak{H}(\mathfrak{Y})}$. Since $\mathfrak{X}$ is closed it follows that $\mathfrak{X} = \overline{\mathfrak{H}(\mathfrak{Y})}$.

EXERCISES 1·8

1. A finite set $\mathfrak{X}$ of N points is given in R^n, $N \geqslant n+2$. Show that it is possible to divide $\mathfrak{X}$ into two subsets $\mathfrak{X}_1$ and $\mathfrak{X}_2$ such that $\mathfrak{H}(\mathfrak{X}_1) \cap \mathfrak{H}(\mathfrak{X}_2) \neq \phi$. (*Hint.* Use projection and induction.)

2. Show that a closed convex set $\mathfrak{X}$ that is neither a linear manifold nor a half-space is the convex cover of its frontier.

3. Show that every exposed point of a convex set is also an extreme point, but (by means of a suitable example) that the converse is not necessarily true.

9. Duality in Euclidean space

A dual space to R^n is a space in which the hyperplanes or half-spaces of R^n are represented by points or, possibly, half-lines. There is no exact duality as in the case of projective geometry, and for that reason there is a choice of dual spaces available. Sometimes it is convenient to use one and sometimes another. They will all be metric spaces and will be topologically equivalent with regard to the convergence properties of a sequence of hyperplanes. The idea of duality as distinct from its precise definition is very important for three reasons. It often suggests alternative proofs of known results; it suggests new results which are 'dual' to known results; it helps to clarify existing knowledge and to coordinate diverse results.

Of the three dual spaces given below we shall use the first only, except that we use $\mathsf{E}(n)$ in § 2·4.

(i) *The polar duality*

Consider two n-dimensional Euclidean spaces R^n and $(\mathsf{R}^n)^*$.

DEFINITION. *If $\mathfrak{X}$ is any subset of R^n, then the polar of $\mathfrak{X}$ is that subset $\mathfrak{Y}$ of $(\mathsf{R}^n)^*$ defined by*

$$\mathfrak{Y}=\{\mathbf{y}:\mathbf{x}.\mathbf{y}\leqslant 1, \mathbf{x}\in\mathfrak{X}\}.$$

If $\mathfrak{X}$ is a single point, then $\mathfrak{Y}$ is a closed half-space, and since the polar of $\bigcup_{i\in I}\mathfrak{X}_i$ is the intersection of the polars of the sets $\mathfrak{X}_i$, it follows that for any $\mathfrak{X}$, $\mathfrak{Y}$ is a closed convex set. Further it is clear that $\mathfrak{Y}\supset\mathbf{O}$ and that if $\mathbf{y}\in\mathfrak{Y}$ so does $\lambda\mathbf{y}$, $0\leqslant\lambda\leqslant 1$.

To indicate the dependence of $\mathfrak{Y}$ upon $\mathfrak{X}$ we shall write $\mathfrak{X}^*$ for $\mathfrak{Y}$.

By proceeding in a similar way we can define the polar of $\mathfrak{X}^*$ as a subset $\mathfrak{X}^{**}$ of a third space $(\mathsf{R}^n)^{**}$. It is customary to identify this space with R^n so that $\mathfrak{X}^{**}$ may be compared directly with $\mathfrak{X}$ and $\mathfrak{X}^{***}$ with $\mathfrak{X}^*$.

The relationship between $\mathfrak{X}^{**}$ and $\mathfrak{X}$ on the one hand and between $\mathfrak{X}^{***}$ and $\mathfrak{X}^*$ on the other is an extremely simple one. We have $\mathfrak{X}^{**}=\mathfrak{H}(\overline{\mathfrak{X}},\mathbf{O})$ and $\mathfrak{X}^{***}=\mathfrak{X}^*$. For if we denote the convex cover of $\overline{\mathfrak{X}}\cup\mathbf{O}$ by $\mathfrak{K}$ then clearly $\mathfrak{K}^*=\mathfrak{X}^*$. Thus to prove the above relations it is sufficient to show that if $\mathfrak{X}$ is any closed convex set that contains $\mathbf{O}$ then $\mathfrak{X}^{**} = \mathfrak{X}$. Now if $\mathbf{x} \in \mathfrak{X}$, then $\mathbf{x}.\mathbf{y} \leqslant 1$ for all $\mathbf{y} \in \mathfrak{X}^*$ and so $\mathbf{x} \in \mathfrak{X}^{**}$; thus $\mathfrak{X}^{**} \supset \mathfrak{X}$; if on the other hand $\mathbf{a} \in \mathsf{R}^n$ does not belong to $\mathfrak{X}$, there is a hyperplane that separates $\mathbf{a}$ strictly from $\mathfrak{X}$. This hyperplane does not go through the origin since the origin lies in $\mathfrak{X}$. Thus we can find $\mathbf{y} \in (\mathsf{R}^n)^*$ such that

$$\mathbf{x}.\mathbf{y} < 1, \mathbf{x} \in \mathfrak{X}; \quad \mathbf{a}.\mathbf{y} > 1.$$

Hence $\mathbf{y} \in \mathfrak{X}^*$ and thus $\mathbf{a} \notin \mathfrak{X}^{**}$. Thus $\mathfrak{X}^{**} \subset \mathfrak{X}$ and we have $\mathfrak{X}^{**} = \mathfrak{X}$.

This type of duality is particularly useful when we are dealing with a single convex set. Suppose that $\mathfrak{X}$ is a convex set with interior points of which the origin is one. If $\mathfrak{X}$ is bounded then $\mathfrak{X}^*$ contains the origin of $(\mathsf{R}^n)^*$ as an interior point. For the relation $\mathfrak{X} \subset \mathfrak{S}(\mathbf{O}, R)$ implies that $|\mathbf{x}| < R$ and thus if $|\mathbf{y}| < 1/R$ we have $|\mathbf{x}.\mathbf{y}| < 1$, i.e. $\mathbf{y} \in \mathfrak{X}^*$. If $\mathfrak{X}$ is a closed bounded set with the origin as an interior point, then $\mathfrak{X}^*$ possesses precisely these qualities in $(\mathsf{R}^n)^*$. For the rest of this paragraph we describe such a set $\mathfrak{X}$ as being normal.

It is sometimes convenient to regard this duality as one between points and hyperplanes. If $\mathbf{x} \neq \mathbf{O}$ is a point of R^n then the dual set $\mathbf{x}^*$ in $(\mathsf{R}^n)^*$ is bounded by a hyperplane which may be denoted by $\mathfrak{P}(\mathbf{x})$. We can say that this hyperplane is dual to $\mathbf{x}$. Similarly, if $\mathfrak{P}$ is a hyperplane in R^n its dual in $(\mathsf{R}^n)^*$ is a segment joining $\mathbf{O}$ to a point $\mathbf{x}^*$ of $(\mathsf{R}^n)^*$, unless $\mathfrak{P}$ passes through the origin in R^n in which case its dual is an infinite line. If we exclude this last case then we can say that $\mathbf{x}^*$ is the dual of $\mathfrak{P}$ and write it as $\mathbf{x}^*(\mathfrak{P})$.

If $\mathfrak{X}$ is normal then the frontier points are dual to support hyperplanes of $\mathfrak{X}^*$ and vice versa. Similarly, the points exterior to $\mathfrak{X}$ are dual to hyperplanes that cut $\mathfrak{X}^*$. For example, the result dual to Theorem 6 is that the class of hyperplanes which are support hyperplanes of a closed bounded set $\mathfrak{X}$ and which pass through a given point exterior to $\mathfrak{X}$ have a point-set union that is non-dense in R^n.

As another example of duality we give some definitions and properties of hyperplanes dual to corresponding properties of points that have already been established.

We shall say that a collection of hyperplanes $\mathfrak{P}_\nu$, $\nu = 1, 2, \ldots, k$, where $\mathfrak{P}_\nu$ has the equation

$$\mathbf{a}^{(\nu)}.\mathbf{x} = a_1^{(\nu)}x_1 + \ldots + a_n^{(\nu)}x_n = 1 \quad (\nu = 1, 2, \ldots, k), \tag{1·9·1}$$

is linearly dependent if there exist real numbers $\lambda_1, \ldots, \lambda_k$ not all zero such that

$$a_i^{(1)}\lambda_1 + \ldots + a_i^{(k)}\lambda_k = 0 \quad (i = 1, 2, \ldots, n),$$
$$\lambda_1 + \ldots + \lambda_k = 0. \tag{1·9·2}$$

Otherwise the hyperplanes are linearly independent. Clearly the hyperplanes (1·9·1) are linearly dependent if and only if the vectors $\mathbf{a}^{(\nu)} - \mathbf{a}^{(1)}$, $\nu = 2, \ldots, k$, of $(\mathsf{R}^n)^*$ are linearly dependent.

If we have $n+1$ linearly independent hyperplanes of the form (1·9·1), then any other hyperplane

$$\mathbf{a}.\mathbf{x} = a_1x_1 + \ldots + a_nx_n = 1$$

can be written as
$$a_i = \sum_{j=1}^{n+1} a_i^{(j)}\lambda_j,$$
$$1 = \sum_{j=1}^{n+1} \lambda_j.$$

The dual of the definition of an extreme point of a convex set is that of an extreme support hyperplane, which is as follows:

Definition. *A hyperplane* $\mathfrak{P}$ $\mathbf{a}.\mathbf{x} = 1$ *is said to be an extreme hyperplane of a closed bounded convex set* $\mathfrak{X}$ *if it does not cut* $\mathfrak{X}$ *and it is not possible to find two hyperplanes* $\mathfrak{P}_1$, $\mathfrak{P}_2$, *say* $\mathbf{a}_1.\mathbf{x} = 1$ *and* $\mathbf{a}_2.\mathbf{x} = 1$, *which do not cut* $\mathfrak{X}$ *and are such that* $\mathbf{a} = \lambda\mathbf{a}_1 + \mu\mathbf{a}_2$, $\lambda + \mu = 1, \lambda > 0, \mu > 0$.

We outline proofs of the following properties:

(i) Every point on the frontier of $\mathfrak{X}$ lies on at least one extreme support hyperplane.

(ii) A closed bounded convex set $\mathfrak{X}$ with interior points is the intersection of the closed half-spaces bounded by its extreme support hyperplanes (dual of Theorem 13).

Proof of (i). Let $\mathbf{x}$ belong to the frontier of $\mathfrak{X}$ and $\mathfrak{C}$ be the closure of the cone subtended by $\mathfrak{X}$ at $\mathbf{x}$.† Let $\mathfrak{P}$ be a hyperplane

† Strictly speaking 'subtended by the relative interior of $\mathfrak{X}$ at $\mathbf{x}$'.

that meets $\mathfrak{C}$ and does not contain $\mathbf{x}$. If an extreme support $(n-2)$-dimensional linear manifold $\mathfrak{Q}$ to $\mathfrak{P} \cap \mathfrak{C}$ lying in $\mathfrak{P}$ meets $\mathfrak{P} \cap \mathfrak{C}$ in say $\mathbf{y}$, then the $(n-1)$-dimensional linear manifold spanned by this space $\mathfrak{Q}$ and the line $\mathbf{xy}$ is an extreme support hyperplane to $\mathfrak{X}$.

The proof is now completed by an inductive argument.

Proof of (ii). If $\mathbf{y} \notin \mathfrak{X}$ and $\mathbf{x}_i \in \mathfrak{X}^0$, $\mathfrak{H}(\mathbf{y}, \mathbf{x}_i)$ meets $\operatorname{Fr} \mathfrak{X}$ in, say, $\mathbf{x}_f$. By (i) $\mathbf{x}_f$ lies on an extreme support hyperplane of $\mathfrak{X}$, say $\mathfrak{P}$. Since $\mathfrak{P}$ separates $\mathbf{y}$ from $\mathbf{x}_i$, $\mathbf{y}$ does not belong to the set $\mathfrak{R}$ which is the intersection of closed half-spaces bounded by extreme support hyperplanes. Thus $\mathfrak{R} \subset \mathfrak{X}$; since the inclusion $\mathfrak{X} \subset \mathfrak{R}$ is trivial we have the required result.

(ii) *The dual space* $\mathsf{E}(n)$

To each hyperplane

$$a_1 x_1 + \ldots + a_n x_n = b \quad (b \geqslant 0) \qquad (1\cdot9\cdot3)$$

make correspond the point $(a_1, a_2, \ldots, a_n, b)$ in an $n+1$-dimensional Euclidean space $\mathsf{E}(n)$. Thus to a single hyperplane in R^n there corresponds a half-ray, terminating at the origin, in $\mathsf{E}(n)$.

(iii) *The dual space* $\mathsf{S}(n)$

To each hyperplane (1·9·3) make correspond the point $(\lambda a_1, \ldots, \lambda a_n, \lambda b)$, where $\lambda = 1/(a_1^2 + \ldots + a_n^2 + b^2)^{\frac{1}{2}}$. Thus to each hyperplane in R^n there corresponds a unique point in an $n+1$-dimensional Euclidean space. This point lies on the surface of the unit sphere which we denote by $\mathsf{S}(n)$.

Exercises 1·9

1. Show that if $\mathfrak{X}$ is a two-dimensional convex set, then every point on the frontier of $\mathfrak{X}$ lies on at least one extreme support hyperplane.

2. Extend the result of Exercise 1·9 (1) to n dimensions by an inductive argument.

3. For each point $\mathbf{x}$ of a closed bounded convex set $\mathfrak{X}$, let $\mathfrak{P}(\mathbf{x})$ be the intersection of all the support hyperplanes of $\mathfrak{X}$ at $\mathbf{x}$ and $\mathfrak{C}(\mathbf{x}) = \mathfrak{P}(\mathbf{x}) \cap \mathfrak{X}$. Let $\mathfrak{R}(\mathbf{x})$ be the set of all points $\mathbf{y}$ such that the line $\mathbf{xy}$ meets $\mathfrak{X}$ in a segment of which $\mathbf{x}$ is a relatively interior point. Show that if $\mathbf{x} \in \operatorname{Fr} \mathfrak{X}$ then $\mathfrak{C}(\mathbf{x}) \supset \mathfrak{R}(\mathbf{x})$.

10. Convex polytopes

The study of convex polytopes has received a considerable impetus from its applications to mathematical economics and to the theory of games. We give here only those results which are required later.

DEFINITION. *A convex polytope is a set which is the convex cover of a finite number of points.*

THEOREM 14. *The convex cover of the points* $\mathbf{x}_1, \mathbf{x}_2, \ldots, \mathbf{x}_k$ *is identical with the set of points of the form*

$$\mathbf{x} = \lambda_1 \mathbf{x}_1 + \ldots + \lambda_k \mathbf{x}_k \quad (\lambda_1 + \ldots + \lambda_k = 1;\ \lambda_i \geqslant 0;\ i = 1, 2, \ldots, k). \tag{1·10·1}$$

Denote the set of points of the form (1·10·1) by $\mathfrak{X}$. It is easily verified that $\mathfrak{X}$ is convex and as $\mathfrak{X} \supset \mathbf{x}_i$, $i = 1, \ldots, k$, it follows that $\mathfrak{X} \supset \mathfrak{H}(\mathbf{x}_1, \ldots, \mathbf{x}_k)$. On the other hand, by (1·2·2) any convex set that contains $\mathbf{x}_1, \ldots, \mathbf{x}_k$ also contains $\mathfrak{X}$. Thus $\mathfrak{X} \subset \mathfrak{H}(\mathbf{x}_1, \ldots, \mathbf{x}_k)$ and the theorem is proved.

THEOREM 15. *If the set* $\mathfrak{X}$ *is the non-void intersection of a finite number of closed half-spaces and is bounded, then* $\mathfrak{X}$ *is a convex polytope.*

By Theorem 13 it is sufficient to show that $\mathfrak{X}$ has a finite number of extreme points.

When $n = 1$ $\mathfrak{X}$ is either a single point or a closed segment and thus has a finite number of extreme points.

Assume inductively that a bounded intersection of a finite number of half-spaces in an $(n-1)$-dimensional space has at most a finite number of extreme points. The extreme points of $\mathfrak{X}$ belong to the frontier of $\mathfrak{X}$ and thus lie on one or other of the hyperplanes which bound the half-spaces of which $\mathfrak{X}$ is the intersection. Denote these hyperplanes by $\mathfrak{P}_1, \ldots, \mathfrak{P}_l$. An extreme point of $\mathfrak{X}$ in $\mathfrak{P}_i$ is an extreme point of $\mathfrak{P}_i \cap \mathfrak{X}$ (cf. the proof of Theorem 13), and there are a finite number of such points by the induction hypothesis. Thus there are a finite number of extreme points of $\mathfrak{X}$ and the theorem is proved.

The converse of Theorem 15 is also true.

THEOREM 16. *If $\mathfrak{X}$ is a convex polytope then it is the intersection of a finite number of closed half-spaces.*

Suppose that $\mathfrak{X}$ is the convex cover of the points $\mathbf{x}_1, \mathbf{x}_2, \ldots, \mathbf{x}_k$, and suppose that it is not the convex cover of any proper subset of $\mathbf{x}_1, \ldots, \mathbf{x}_k$. $\mathfrak{X}$ is the intersection of the closed half-spaces bounded by extreme support hyperplanes. Thus we have only to show that there are at most a finite number of extreme support hyperplanes to $\mathfrak{X}$.

Assume inductively that this is true in space of $n-1$ dimensions. Project $\mathfrak{X}$ from $\mathbf{x}_k$ onto a hyperplane $\mathfrak{P}$. We can take $\mathfrak{P}$ to be parallel to a support hyperplane to $\mathfrak{X}$ at $\mathbf{x}_k$ that meets $\mathfrak{X}$ only in the point $\mathbf{x}_k$. Denote this projection of $\mathfrak{X}$ by $\mathfrak{X}_1$. $\mathfrak{X}_1$ is the convex cover of the projections of $\mathbf{x}_1, \ldots, \mathbf{x}_{k-1}$, and thus there are at most a finite number of extreme support $(n-2)$-dimensional linear manifolds to $\mathfrak{X}_1$ in $\mathfrak{P}$. Now any extreme support hyperplane to $\mathfrak{X}$ that passes through $\mathbf{x}_k$ intersects $\mathfrak{P}$ in an $(n-2)$-dimensional linear manifold which is an extreme support space of $\mathfrak{X}_1$ in $\mathfrak{P}$. Thus there are at most a finite number of extreme support hyperplanes to $\mathfrak{X}$ that pass through $\mathbf{x}_k$. The same is true of each of the points $\mathbf{x}_1, \ldots, \mathbf{x}_{k-1}$, and since every support hyperplane to $\mathfrak{X}$ passes through an extreme point of $\mathfrak{X}$, i.e. one of $\mathbf{x}_i$, $i = 1, 2, \ldots, k$, it follows that there are at most a finite number of extreme support hyperplanes.

This completes the proof of the theorem.

These two alternative definitions of convex polytopes as either (*a*) the convex cover of a finite number of points, or (*b*) the non-void bounded intersection of a finite number of closed half-spaces, illustrate the scope and the limitations of the concept of duality in Euclidean space. The conditions of being non-void and bounded in (*b*) cannot be dispensed with, but the *essential* properties that are dual are those of being the intersection of a finite number of closed half-spaces on the one hand or of being the convex cover of a finite set of points on the other. An examination of Theorems 15 and 16 will show that we have used arguments that are essentially duals of one another.

11. Continuous mappings of convex sets. Regular convex sets

It is sometimes convenient to reduce a particular problem concerning a closed bounded convex set $\mathfrak{X}$ to a problem on a sphere by mapping $\mathfrak{X}$ onto a sphere. There are essentially two methods of doing this, and we describe these methods in this paragraph.

(*a*) If $\mathfrak{X}$ is an r-dimensional closed bounded convex set and $\mathbf{x}_0$ is a relative interior point of $\mathfrak{X}$, then there is a homeomorphism $\mathscr{T}$ of the whole space R^n onto itself such that $\mathscr{T}(\mathfrak{X})$ is a solid r-dimensional sphere and $\mathscr{T}(\mathbf{x}_0)$ is its centre.

It is sufficient to consider the case when $\mathfrak{X}$ is n-dimensional. Let $\mathbf{x}$ be a point of R^n. The half-ray of the line $\mathbf{x}\mathbf{x}_0$ that terminates at $\mathbf{x}_0$ and contains $\mathbf{x}$ meets the frontier of $\mathfrak{X}$ in the unique point $f(\mathbf{x})$ provided $\mathbf{x} \neq \mathbf{x}_0$. Define the transformation $\mathscr{T}$ by

$$\mathscr{T}(\mathbf{x}) = (\mathbf{x}-\mathbf{x}_0)/|f(\mathbf{x})-\mathbf{x}_0| + \mathbf{x}_0 \quad (\mathbf{x} \neq \mathbf{x}_0),$$

$$\mathscr{T}(\mathbf{x}_0) = \mathbf{x}_0.$$

This transformation has the required properties.

To explain the second mapping we need to introduce the idea of regularity.

DEFINITION. (i) *A frontier point of a convex set is regular if it lies on only one support hyperplane.*

(ii) *A support hyperplane of a convex set is regular if it meets the set in only one point.*

(iii) *A convex set is regular if all its frontier points and all its support hyperplanes are regular.*

(*b*) The second type of mapping is a relation between the surface of an n-dimensional, closed bounded convex set $\mathfrak{X}$ and the surface of an n-dimensional sphere. Let $\mathbf{x}$ be a point of the frontier of $\mathfrak{X}$ and let $\Omega_{\mathbf{x}}$ be the class of support hyperplanes of $\mathfrak{X}$ at $\mathbf{x}$. For $\mathfrak{P} \in \Omega_{\mathbf{x}}$ let $\mathfrak{Q}(\mathfrak{P})$ be a hyperplane which is parallel to $\mathfrak{P}$, which touches a fixed n-dimensional closed sphere $\mathfrak{S}$ and which is such that $\mathfrak{S}$ lies on the same side of $\mathfrak{Q}(\mathfrak{P})$ as does $\mathfrak{X}$ of $\mathfrak{P}$. Write

$$\mathbf{s}_{\mathfrak{P}} = \mathfrak{Q}(\mathfrak{P}) \cap \mathfrak{S}.$$

The correspondence in which we are interested is defined by

$$\mathbf{x} \to \bigcup_{\mathfrak{P} \in \Omega_{\mathbf{x}}} \mathbf{s}_{\mathfrak{P}}, \quad \mathbf{s}_{\mathfrak{P}} \to \bigcup_{\mathfrak{P} \in \Omega_{\mathbf{x}}} \mathbf{x}, \tag{1·11·1}$$

and is called the *mapping by parallel support hyperplanes*. In general this correspondence is many-many. If either side is a single point then the correspondence is continuous in the single-valued sense. Thus we have

(i) if every frontier point of $\mathfrak{X}$ is regular then the frontier of $\mathfrak{X}$ is mapped onto the frontier of $\mathfrak{S}$ continuously by the mapping by parallel support hyperplanes,

(ii) if every support hyperplane of $\mathfrak{X}$ is regular then the frontier of $\mathfrak{S}$ is mapped onto the frontier of $\mathfrak{X}$ continuously by the mapping by parallel support hyperplanes.

The proofs are left to the reader.

It is easy to see that any closed bounded n-dimensional convex set $\mathfrak{X}$ with a non-empty interior is such that its set of regular frontier points is dense in the frontier of $\mathfrak{X}$. For if $\mathbf{p}$ is any point of the frontier of $\mathfrak{X}$ and δ is a given positive number we may choose a point $\mathbf{q}$ of the interior of $\mathfrak{X}$ such that $|\mathbf{p}-\mathbf{q}| < \frac{1}{3}\delta$. There is a largest closed sphere $\mathfrak{S}$ whose centre is $\mathbf{q}$ and which is contained in $\mathfrak{X}$ and there is at least one point say $\mathbf{r}$ that belongs to $\mathrm{Fr}\,\mathfrak{X} \cap \mathfrak{S}$. Now $\mathfrak{S}$ cannot have radius greater than $\frac{1}{3}\delta$ or it would include $\mathbf{p}$ as an interior point and we should not have $\mathfrak{S} \subset \mathfrak{X}$. Thus $|\mathbf{p}-\mathbf{r}| < \delta$. But $\mathbf{r}$ is a regular point of $\mathrm{Fr}\,\mathfrak{X}$, for of any two distinct hyperplanes through $\mathbf{r}$ one must cut $\mathfrak{S}$ and hence must cut $\mathfrak{X}$. Thus there is a unique support hyperplane to $\mathfrak{X}$ at $\mathbf{r}$. Since $\mathbf{p}$ is any point of $\mathrm{Fr}\,\mathfrak{X}$ and δ is any positive number this is the required result.

Exercise 1·11

1. Show that to every closed bounded n-dimensional convex set there corresponds at least one regular support hyperplane.

CHAPTER 2

HELLY'S THEOREM AND ITS APPLICATIONS

One of the most striking properties of Euclidean n-dimensional space is a result on the intersection of convex sets due to Helly. This property is closely related to Carathéodory's theorem on the convex cover of a given set, and the relationship is connected with duality. Carathéodory's theorem implies Helly's theorem, and conversely also Helly's theorem implies the dual of Carathéodory's. Here of course we are using the concept of duality in a descriptive and imprecise sense.

The properties of convex sets which were developed in Chapter 1 are true in one form or another in Banach spaces of either finite or infinite dimension. This is no longer the case with the theorems that are to be proved in the present chapter. A vector space which satisfies Helly's theorem is essentially one whose dimension is finite. It is possible to generalize Helly's theorem by a process of axiomatization, but we shall not do so here.

1. Radon's proof of Helly's theorem

We give here a simple analytical proof of Helly's theorem due to Radon.

THEOREM 17. HELLY'S THEOREM. *A finite class of N convex sets in R^n is such that $N \geqslant n+1$, and to every subclass which contains $n+1$ members there corresponds a point of R^n which belongs to every member of the subclass. Under these conditions there is a point which belongs to every member of the given class of N convex sets.*

The theorem is trivially true when $N = n+1$. Assume inductively that the theorem is true for every class of $N-1$ sets and consider the case when the class contains N sets.

Let the sets be $\mathfrak{X}_1, \ldots, \mathfrak{X}_N$. By the inductive hypothesis applied to the subclass of this class of sets which consists of the whole class except $\mathfrak{X}_i$ there is a point $\mathbf{x}^{(i)}=(x_1^{(i)}, \ldots, x_n^{(i)})$ which belongs to $\mathfrak{X}_j$ provided $j \neq i$.

The equations

$$\left.\begin{aligned} \sum_{i=1}^{N} \lambda_i x_k^{(i)} &= 0 \quad (k=1, 2, \ldots, n), \\ \sum_{i=1}^{N} \lambda_i \quad &= 0, \end{aligned}\right\} \qquad (2\cdot1\cdot1)$$

form a set of $n+1$ equations in the N unknowns $\lambda_1, \lambda_2, \ldots, \lambda_N$. Since $N > n+1$ these equations have non-trivial sets of solutions. For one such solution denote by $\lambda_{j_1}, \ldots, \lambda_{j_k}$ those of the λ_i that are non-negative and by $\lambda_{h_1}, \ldots, \lambda_{h_{N-k}}$ those that are negative. Define the point $\mathbf{y}=(y_1, \ldots, y_n)$ by

$$y_k = (\sum_r \lambda_{j_r} x_k^{(j_r)})/(\sum_r \lambda_{j_r}). \qquad (2\cdot1\cdot2)$$

Since $\mathbf{x}^{(j_r)} \in \mathfrak{X}_j$ provided $j \neq j_r$, it follows from (1·2·2) that $\mathbf{y} \in \mathfrak{X}_j$ provided $j \neq j_1, \ldots, j_k$, i.e. $\mathbf{y}$ belongs to each of $\mathfrak{X}_{h_1}, \mathfrak{X}_{h_2}, \ldots, \mathfrak{X}_{h_{N-k}}$. But the equations (2·1·1) enable us to write (2·1·2) in the form

$$y_k = (\sum_s (-\lambda_{h_s}) x_k^{(h_s)})/(\sum_s -\lambda_{h_s}), \qquad (2\cdot1\cdot3)$$

and this shows similarly that $\mathbf{y}$ belongs to $\mathfrak{X}_{j_1}, \mathfrak{X}_{j_2}, \ldots, \mathfrak{X}_{j_k}$. Thus $\mathbf{y}$ is a point common to all the sets $\mathfrak{X}_i$, $i=1, \ldots, N$.

Exercises 2·1

1. Use Exercise 1·8·1 to obtain an alternative proof of Helly's theorem.

2. A class of convex sets in R^n is such that to every subclass of $n+1$ members of the class there corresponds a point whose distance from each of the $n+1$ convex sets is less than or equal to a fixed positive number d. Show that there is a point whose distance from each convex set of the whole class is less than or equal to d.

2. Carathéodory's theorem

In this paragraph we prove the theorem of Carathéodory and an extension of it. This theorem may be described by saying that the convex cover of a given set $\mathfrak{X}$ is the union of an aggregate of simplexes whose vertices are amongst the points $\mathfrak{X}$.

Carathédory's theorem may be proved geometrically using Theorem 10 and as we shall see later we may deduce Theorem 10 from Carathéodory's theorem. The exact logical status of this theorem and of Helly's theorem is difficult to determine. Geometrically they seem to be the duals of each other, yet analytically they can be proved by what is effectively the same sort of argument. There is also an extension of Carathéodory's theorem in which we assume that the set $\mathfrak{X}$ in R^n has at most n components.

THEOREM 18. (i) *If* $\mathbf{y}$ *is a point of* $\mathfrak{H}(\mathfrak{X})$ *there is a set of s points* $\mathbf{x}_1, \dots, \mathbf{x}_s$ *all belonging to* $\mathfrak{X}$ *with* $s \leqslant n+1$ *such that* $\mathbf{y}$ *is a point of the simplex whose vertices are* $\mathbf{x}_1, \dots, \mathbf{x}_s$.

(ii) *If, in addition,* $\mathfrak{X}$ *has at most n components then the statement of* (i) *is true with* $s \leqslant n$.

We first establish a subsidiary result.

LEMMA. *The set of points contained in* R^n *of the form*

$$\lambda_1 \mathbf{x}_1 + \dots + \lambda_k \mathbf{x}_k$$

where k is any positive integer, $\mathbf{x}_1, \dots, \mathbf{x}_k$ *are any points of* $\mathfrak{X}$ *and* $\lambda_1, \dots, \lambda_k$ *are numbers satisfying*

$$\lambda_1 + \lambda_2 + \dots + \lambda_k = 1, \quad \lambda_i \geqslant 0 \quad (i = 1, 2, \dots, k),$$

coincides with the set $\mathfrak{H}(\mathfrak{X})$.

Denote the set described in the lemma by $\mathfrak{K}$. First for any particular point $\mathbf{x}$ of $\mathfrak{K}$,

$$\mathbf{x} = \lambda_1 \mathbf{x}_1 + \lambda_2 \mathbf{x}_2 + \dots + \lambda_k \mathbf{x}_k$$

for some positive integer k and thus, by § 1·2, $x \in \mathfrak{H}(\mathbf{x}_1, \dots, \mathbf{x}_k)$. But $\mathbf{x}_1, \dots, \mathbf{x}_k \in \mathfrak{X}$ and thus

$$\mathfrak{H}(\mathbf{x}_1, \dots, \mathbf{x}_k) \subset \mathfrak{H}(\mathfrak{X}).$$

Hence $\mathbf{x} \in \mathfrak{H}(\mathfrak{X})$ and since this is true for every $\mathbf{x} \in \mathfrak{K}$ it follows that

$$\mathfrak{K} \subset \mathfrak{H}(\mathfrak{X}).$$

On the other hand the set $\mathfrak{K}$ is convex. For suppose that $\mathbf{y} \in \mathfrak{K}$ and $\mathbf{z} \in \mathfrak{K}$ where

$$\mathbf{y} = \lambda_1 \mathbf{y}_1 + \dots + \lambda_s \mathbf{y}_s,$$
$$\mathbf{z} = \mu_1 \mathbf{z}_1 + \dots + \mu_t \mathbf{z}_t.$$

Consider the point $\mathbf{x} \in \mathfrak{H}(\mathbf{y}, \mathbf{z})$ where

$$\mathbf{x} = \chi \mathbf{y} + (1 - \chi) \mathbf{z} \quad (0 \leqslant \chi \leqslant 1).$$

Then $\mathbf{x} = \chi\lambda_1\mathbf{y}_1 + \ldots + \chi\lambda_s\mathbf{y}_s + (1-\chi)\mu_1\mathbf{z}_1 + \ldots + (1-\chi)\mu_t\mathbf{z}_t$

and since $\chi\lambda_i \geqslant 0$, $(1-\chi)\mu_j \geqslant 0$ and

$$\chi\lambda_1 + \ldots + \chi\lambda_s + (1-\chi)\mu_1 + \ldots + (1-\chi)\mu_t = 1$$

it follows that $\mathbf{x} \in \mathfrak{K}$.

Clearly
$$\mathfrak{X} \subset \mathfrak{K}$$
and therefore, since $\mathfrak{K}$ is convex
$$\mathfrak{H}(\mathfrak{X}) \subset \mathfrak{K}.$$
Hence finally $\mathfrak{K} = \mathfrak{H}(\mathfrak{X})$ and the lemma is proved.

(i) We next prove (i) of the theorem.

Suppose that $\mathbf{x} \in \mathfrak{H}(\mathfrak{X})$. By the lemma we know that there is some finite integer k such that

$$\mathbf{x} = \lambda_1\mathbf{x}_1 + \ldots + \lambda_k\mathbf{x}_k \quad (\mathbf{x}_i \in \mathfrak{X}), \tag{2·2·1}$$

where
$$\lambda_1 + \ldots + \lambda_k = 1, \quad \lambda_i \geqslant 0 \quad (i = 1, 2, \ldots, k).$$

Our aim is to show that we can find such an expression for $\mathbf{x}$ with $k \leqslant n+1$. The k points $\mathbf{x}_1, \ldots, \mathbf{x}_k$, are linearly dependent if $k > n+1$ and we can find numbers $\alpha_1, \ldots, \alpha_k$, not all zero such that

$$\alpha_1\mathbf{x}_1 + \ldots + \alpha_k\mathbf{x}_k = \mathbf{O}, \quad \alpha_1 + \ldots + \alpha_k = 0.$$

The set of real numbers τ for which

$$\tau\alpha_i \geqslant -\lambda_i, \quad (i = 1, 2, \ldots, k)$$

is closed, is non-void (since it contains the number O), and does not contain all the real axis (since at least one α_i is non-zero). Let τ_0 be a frontier point of this set. Then

$$\mathbf{x} = (\lambda_1 + \tau_0\alpha_1)\mathbf{x}_1 + \ldots + (\lambda_k + \tau_0\alpha_k)\mathbf{x}_k,$$

and
$$\lambda_i + \tau_0\alpha_i \geqslant 0 \quad (i = 1, 2, \ldots, k),$$
$$(\lambda_1 + \tau_0\alpha_1) + \ldots + (\lambda_k + \tau_0\alpha_k) = 1.$$

Further for at least one integer i, $1 \leqslant i \leqslant k$

$$\lambda_i + \tau_0\alpha_i = 0$$

and we have an expression for $\mathbf{x}$ of the same form as (2·2·1) except that, since one of the coefficients is zero, there are only $k-1$ significant terms.

The process may be repeated until we have expressed $\mathbf{x}$ as a linear combination of at most $n+1$ of the points of $\mathfrak{X}$.

This completes the proof of (i) of the theorem.

Proof of (ii). This depends upon (i). Let $\mathbf{y}$ be any point of $\mathfrak{H}(\mathfrak{X})$. There are s points of $\mathfrak{X}$ with $s \leqslant n+1$, $\mathbf{y} \in \mathfrak{H}(\mathbf{x}_1, \ldots, \mathbf{x}_s)$. If $s < n+1$ (ii) is true. Suppose then that $s = n+1$. Write

$$\mathbf{y} = \lambda_1 \mathbf{x}_1 + \ldots + \lambda_{n+1} \mathbf{x}_{n+1} \quad (\lambda_i \geqslant 0, \lambda_1 + \ldots + \lambda_{n+1} = 1), \quad (2\cdot2\cdot2)$$

where if any $\lambda_i = 0$ (ii) is true, so we may suppose

$$\lambda_i > 0 \quad (i = 1, \ldots, n+1).$$

Let $\mathbf{x}'_i$ be the reflexion of $\mathbf{x}_i$ in $\mathbf{y}$, i.e.

$$\mathbf{x}'_i = 2\mathbf{y} - \mathbf{x}_i \quad (1 \leqslant i \leqslant n+1). \quad (2\cdot2\cdot3)$$

Let $\mathfrak{C}_j$ be the cone subtended at $\mathbf{y}$ by the simplex whose vertices are the n points $\mathbf{x}'_1, \mathbf{x}'_2, \ldots, \mathbf{x}'_{j-1}, \mathbf{x}'_{j+1}, \ldots, \mathbf{x}'_{n+1}$ (see fig. 5 for the

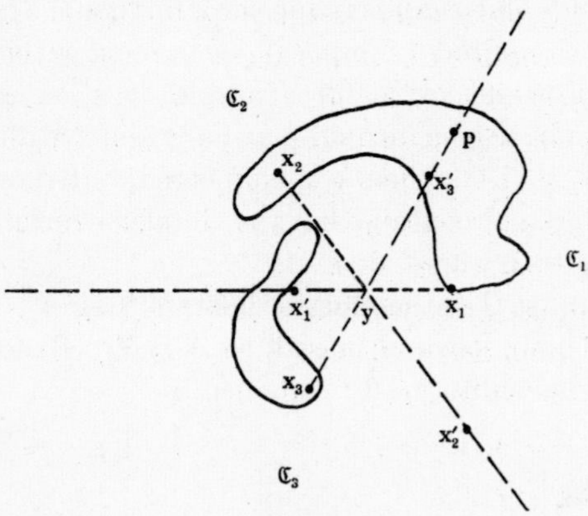

Fig. 5

case $n = 2$). We consider for definiteness the cone $\mathfrak{C}_1$; its points $\mathbf{z}$ are of the form

$$\mathbf{z} = \chi\mathbf{y} + \tau(\mu_2 \mathbf{x}'_2 + \ldots + \mu_{n+1} \mathbf{x}'_{n+1}),$$

where $\mu_i \geqslant 0$, $\mu_2 + \ldots + \mu_{n+1} = 1$, $\chi + \tau = 1$, $\tau \geqslant 0$. Take $\mathbf{y}$ to be the origin; then

$$\mathbf{z} = \mu'_2 \mathbf{x}'_2 + \ldots + \mu'_{n+1} \mathbf{x}'_{n+1} \quad (\mu'_i \geqslant 0, 2 \leqslant i \leqslant n+1), \quad (2\cdot2\cdot4)$$

and similarly for the sets $\mathfrak{C}_j$, $2 \leqslant j \leqslant n+1$.

By (2·2·2) and (2·2·3)

$$\mathbf{x}'_j = -\mathbf{x}_j = \frac{1}{\lambda_j}(\lambda_1 \mathbf{x}_1 + \ldots + \lambda_{j-1}\mathbf{x}_{j-1} + \lambda_{j+1}\mathbf{x}_{j+1} + \ldots + \lambda_{n+1}\mathbf{x}_{n+1}).$$

Thus $\mathbf{x}_j$ is an interior point of $\mathfrak{C}_j$. Since there are $n+1$ cones each containing points of $\mathfrak{X}$, and since $\mathfrak{X}$ has at most n components, there is at least one point of $\mathfrak{X}$ in the frontier of one of the cones $\mathfrak{C}_j$. Suppose, for example, that $\mathbf{p} \in \mathfrak{X} \cap \operatorname{Fr} \mathfrak{C}_1$. Now $\operatorname{Fr} \mathfrak{C}_1$ is formed by points $\mathbf{z}$ of the form (2·2·4) in which at least one of the numbers $\mu_2', \ldots, \mu_{n+1}'$ is zero. Suppose then that $\mu_2' = 0$, say

$$\mathbf{p} = \chi_3 \mathbf{x}_3' + \ldots + \chi_{n+1} \mathbf{x}_{n+1}' \quad (\chi_i \geqslant 0;\ i = 3, \ldots, n+1).$$

Then

$$\mathbf{y} = \mathbf{O} = (\mathbf{p} + \chi_3 \mathbf{x}_3 + \ldots + \chi_{n+1} \mathbf{x}_{n+1})/(1 + \chi_3 + \ldots + \chi_{n+1}).$$

Thus $\mathbf{y}$ is a point of the simplex whose n vertices $\mathbf{p}, \mathbf{x}_3, \ldots, \mathbf{x}_{n+1}$ all belong to $\mathfrak{X}$.

The other possible cases are all similar and the proof of (ii) is complete.

Remarks. (i) One cannot reduce any further the least number of points of $\mathfrak{X}$ required by imposing even more severe conditions on the connectivity of $\mathfrak{X}$; for example, if $\mathbf{x}_1, \ldots, \mathbf{x}_{n+1}$ are the vertices of a non-degenerate simplex in R^n and $\mathfrak{X}$ is the n segments $\mathfrak{H}(\mathbf{x}_1, \mathbf{x}_i)$, $2 \leqslant i \leqslant n+1$, then $\mathfrak{X}$ is connected, but there are points of $\mathfrak{H}(\mathfrak{X})$ that are not contained in any simplex which has s vertices all belonging to $\mathfrak{X}$ where $s \leqslant n-1$.

(ii) We can use Carathéodory's theorem to prove Theorem 10, that if $\mathfrak{X}$ is bounded and closed, so is $\mathfrak{H}(\mathfrak{X})$. Denote by $\mathfrak{B}$ the closed bounded subset of R^{n+1} defined by

$$\mathfrak{B} = \{(\lambda_1, \ldots, \lambda_{n+1}) : \lambda_i \geqslant 0,\ 1 \leqslant i \leqslant n+1,\ \lambda_1 + \ldots + \lambda_{n+1} = 1\}.$$

The mapping

$$(\lambda_1, \ldots, \lambda_{n+1}, x_1^{(1)}, \ldots, x_n^{(1)}, x_1^{(2)}, \ldots \ldots, x_1^{(n+1)}, \ldots, x_n^{(n+1)}) \rightarrow \sum_{i=1}^{n+1} \lambda_i (x_1^{(i)}, x_2^{(i)}, \ldots, x_n^{(i)})$$

is continuous and defined over $\mathsf{R}^{n+1} \times \mathsf{R}^n \times \ldots \times \mathsf{R}^n = \mathsf{R}^{(n+1)^2}$. By Carathéodory's theorem it maps $\mathfrak{B} \times \mathfrak{X} \times \ldots \times \mathfrak{X}$ on $\mathfrak{H}(\mathfrak{X})$ which is therefore both closed and bounded.

Exercises 2·2

1. Show that if $\mathbf{x}$ is a point of the convex cover of the k convex sets $\mathfrak{X}_1, \ldots, \mathfrak{X}_k$, then $\mathbf{x}$ is also a point of a simplex whose vertices are such that at most one belongs to any one set $\mathfrak{X}_i$.

2. Show that a closed bounded convex set is the convex cover of its extreme points. (Compare Theorem 13; note that the set of extreme points need not be closed.)

3. The relation of Helly's theorem to Carathéodory's theorem

We shall show first that Helly's theorem can be deduced from Carathéodory's theorem and then that Carathéodory's may be deduced from Helly's theorem. Then we consider a number of similar and related results.

Deduction of Helly's theorem from Carathéodory's

Suppose that we have a collection of N convex sets satisfying the conditions of Theorem 17 and yet such that there is no point that belongs to them all. We shall show that we are led to a contradiction. Suppose, first, that the sets $\mathfrak{X}_1, \mathfrak{X}_2, \ldots, \mathfrak{X}_N$ are closed and bounded.

Let $\mathbf{x}$ be any point of R^n. We can find a point $\mathbf{x}_0$ such that

$$f(\mathbf{x}) = \max_{1 \leqslant r \leqslant N} \rho(\mathbf{x}, \mathfrak{X}_r) \tag{2·3·1}$$

attains its least value at $\mathbf{x} = \mathbf{x}_0$. Clearly $f(\mathbf{x}_0) > 0$. Amongst the $\mathfrak{X}_r$ there are some which we may suppose for convenience to be $\mathfrak{X}_1, \ldots, \mathfrak{X}_k$ which are such that

$$f(\mathbf{x}_0) = \rho(\mathbf{x}_0, \mathfrak{X}_r) \quad (r = 1, \ldots, k). \tag{2·3·2}$$

Suppose that $\mathbf{x}_r \in \mathfrak{X}_r$ and $|\mathbf{x}_0 - \mathbf{x}_r| = f(\mathbf{x}_0)$, $r = 1, 2, \ldots, k$ (see fig. 6). The points $\mathbf{x}_r$ exist because each $\mathfrak{X}_r$ is closed and bounded; each $\mathbf{x}_r$ is unique because each $\mathfrak{X}_r$ is convex. Then

$$\mathbf{x}_0 \in \mathfrak{H}(\mathbf{x}_1, \ldots, \mathbf{x}_k),$$

for otherwise we could reduce $f(\mathbf{x}_0)$ by moving $\mathbf{x}_0$ towards $\mathfrak{H}(\mathbf{x}_1, \ldots, \mathbf{x}_k)$. By Carathéodory's theorem there is a subset of the points $\mathbf{x}_1, \ldots, \mathbf{x}_k$ consisting of at most $n+1$ members, say $\mathbf{x}_1, \ldots, \mathbf{x}_s$, such that $\mathbf{x}_0 \in \mathfrak{H}(\mathbf{x}_1, \mathbf{x}_2, \ldots, \mathbf{x}_s)$. We have

$$\mathbf{x}_0 = \sum_{r=1}^{s} \mu_r \mathbf{x}_r \quad (\mu_1 + \ldots + \mu_s = 1, \mu_i \geqslant 0).$$

The hyperplane through $\mathbf{x}_r$ perpendicular to $\mathbf{x}_0\mathbf{x}_r$ supports $\mathfrak{X}_r$ at $\mathbf{x}_r$. Thus if $\mathbf{y} \in \mathfrak{X}_r$ then

$$(\mathbf{y} - \mathbf{x}_0) . (\mathbf{x}_r - \mathbf{x}_0) \geqslant (|\mathbf{x}_0 - \mathbf{x}_r|)^2 > 0 \quad (r = 1, \ldots, k), \tag{2·3·4}$$

since none of the $\mathbf{x}_r$ coincide with $\mathbf{x}_0, r=1, \ldots, k$. Thus if $\mathbf{y} \in \mathfrak{X}_r$, $r=1, \ldots, s$ (such a point exists since $s \leqslant n+1$ by the hypotheses of Helly's theorem),

$$\begin{aligned} 0 &= (\mathbf{y}-\mathbf{x}_0).(\mathbf{x}_0-\mathbf{x}_0) \\ &= (\mathbf{y}-\mathbf{x}_0).\left(\sum_{r=1}^{s} \mu_r(\mathbf{x}_r-\mathbf{x}_0)\right) \\ &= \sum_{r=1}^{s} \mu_r(\mathbf{y}-\mathbf{x}_0).(\mathbf{x}_r-\mathbf{x}_0) \\ &> 0, \end{aligned}$$

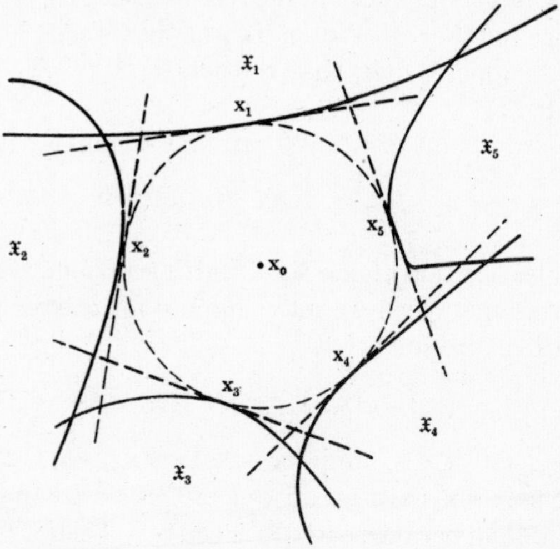

Fig. 6

since all $\mu_r \geqslant 0$ and at least one $\mu_r > 0$. This contradiction establishes Helly's theorem in the case when the sets are closed and bounded. The more general case can now be deduced.

Deduction of Carathéodory's theorem from Helly's theorem

We shall only prove the case of Carathéodory's theorem in which the given set $\mathfrak{X}$ is closed and bounded. The more general result can be deduced from this.

Let $\mathbf{y} \in \mathfrak{H}(\mathfrak{X})$. We can suppose that $\mathbf{y}$ is an interior point of $\mathfrak{H}(\mathfrak{X})$, for if this is not the case $\mathfrak{H}(\mathfrak{X})$ is supported at $\mathbf{y}$ by a hyperplane, say $\mathfrak{P}$, and $\mathbf{y} \in \mathfrak{H}(\mathfrak{X} \cap \mathfrak{P})$. If $\mathbf{y}$ is an interior point of $\mathfrak{H}(\mathfrak{X} \cap \mathfrak{P})$

then we can argue with $\mathfrak{X} \cap \mathfrak{P}$ instead of with $\mathfrak{X}$. Otherwise we repeat the process until we reach a linear manifold $\mathfrak{Q}$ such that $\mathbf{y}$ is an interior point of $\mathfrak{H}(\mathfrak{X} \cap \mathfrak{Q})$ relative to $\mathfrak{Q}$. We suppose then that $\mathbf{y}$ is an interior point of $\mathfrak{H}(\mathfrak{X})$.

If $\mathbf{y} \in \mathfrak{X}$ there is nothing to prove. If $\mathbf{y} \notin \mathfrak{X}$, then for each point $\mathbf{x} \in \mathfrak{X}$ denote by $\mathfrak{T}(\mathbf{x})$, $\mathfrak{W}(\mathbf{x})$ the closed half-spaces bounded by the hyperplane through $\mathbf{x}$ perpendicular to $\mathbf{xy}$ (see fig. 7); $\mathfrak{W}(\mathbf{x})$ is the half-space which contains $\mathbf{y}$. Now the set $\bigcap_{\mathbf{x} \in \mathfrak{X}} \mathfrak{T}(\mathbf{x})$ is empty. For suppose it contained a point $\mathbf{z}$. Let $\mathfrak{P}$ be the hyperplane through $\mathbf{y}$ perpendicular to $\mathbf{zy}$. There is a point, say $\mathbf{x}_0$, of $\mathfrak{X}$

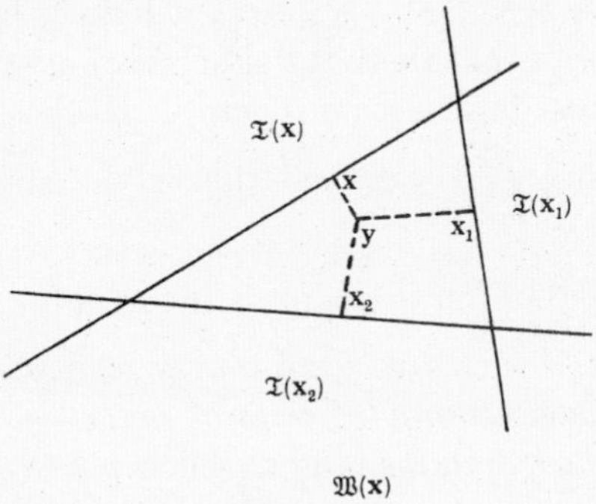

Fig. 7

separated from $\mathbf{z}$ by $\mathfrak{P}$. Then the hyperplane perpendicular to $\mathbf{x}_0\mathbf{y}$ through $\mathbf{x}_0$ does not separate $\mathbf{y}$ from $\mathbf{z}$. A contradiction, since $\mathbf{z} \in \mathfrak{T}(\mathbf{x}_0)$ and $\mathbf{y} \in \mathfrak{W}(\mathbf{x}_0)$. Since each $\mathfrak{T}(\mathbf{x})$ is convex it follows from the converse of Helly's theorem that there are s points, $s \leqslant n+1$, $\mathbf{x}_1, \ldots, \mathbf{x}_s$ of $\mathfrak{X}$ such that

$$\bigcap_{i=1}^{s} \mathfrak{T}(\mathbf{x}_i) = \phi.$$

But this implies that $\mathbf{y} \in \mathfrak{H}(\mathbf{x}_1, \ldots, \mathbf{x}_s)$, for if this is not so there is a hyperplane $\mathfrak{Q}$ strictly separating $\mathbf{y}$ from $\mathfrak{H}(\mathbf{x}_1, \ldots, \mathbf{x}_s)$. Let the half-line that terminates at $\mathbf{y}$, is perpendicular to $\mathfrak{Q}$ and meets $\mathfrak{Q}$, be l. Then l meets every $\mathfrak{T}(\mathbf{x}_i)$ and all of l except a finite segment is contained in $\mathfrak{T}(x_i)$. But this implies that $\bigcap_{i=1}^{s} \mathfrak{T}(\mathbf{x}_i) \neq \phi$. This last relation is false and hence $\mathbf{y} \in \mathfrak{H}(\mathbf{x}_1, \ldots, \mathbf{x}_s)$ and Carathéodory's theorem is proved.

Carathéodory's theorem can also be put in a dual form and we prove next two such forms of the theorem.

THEOREM 19. *A bounded convex set* $\mathfrak{X}$ *is the intersection of a finite number of closed half-spaces* $\mathfrak{J}_1, \ldots, \mathfrak{J}_k$ *bounded by hyperplanes* $\mathfrak{P}_1, \mathfrak{P}_2, \ldots, \mathfrak{P}_k$. *If a hyperplane* $\mathfrak{P}$ *does not meet* $\mathfrak{X}$ *then there is a set of at most n hyperplanes* $\mathfrak{P}_{i_1}, \ldots, \mathfrak{P}_{i_s}$ *from amongst the* $\mathfrak{P}_1, \ldots, \mathfrak{P}_k$ *for which the intersection of the corresponding closed half-spaces bounded by* $\mathfrak{J}_{i_1}, \ldots, \mathfrak{J}_{i_s}$ *forms a set* $\mathfrak{X}_1$ *such that*

$$\mathfrak{X}_1 \supset \mathfrak{X} \quad \textit{and} \quad \mathfrak{P} \cap \mathfrak{X}_1 = \phi.$$

Let $\mathfrak{J}$ denote the closed half-space bounded by $\mathfrak{P}$ that does not contain $\mathfrak{X}$. Since $\mathfrak{J} \cap \mathfrak{J}_1 \cap \ldots \cap \mathfrak{J}_k = \phi$, it follows by the converse of Helly's theorem that there is a subclass of at most $n+1$ of the $\mathfrak{J}, \mathfrak{J}_1, \ldots, \mathfrak{J}_k$ whose intersection is empty. Of these $n+1$ sets one must be $\mathfrak{J}$, since $\bigcap_{i=1}^{k} \mathfrak{J}_i \neq \phi$, and this is the statement of the theorem.

The direct dual of Carathéodory's theorem is

THEOREM 20. *A bounded closed convex set* $\mathfrak{X}$ *with a non-empty interior is the intersection of a class of closed half-spaces. If the hyperplane* $\mathfrak{P}$ *whose equation is* $\mathbf{a}.\mathbf{x} = 1$ *does not meet* $\mathfrak{X}$ *it is possible to find a collection of at most* $n+1$ *hyperplanes* $\mathfrak{P}_1, \ldots, \mathfrak{P}_{n+1}$ *each bounding a closed half-space of the given class such that*

$$\mathbf{a} = \lambda_1 \mathbf{a}_1 + \ldots + \lambda_{n+1} \mathbf{a}_{n+1} \quad (\lambda_1 + \ldots + \lambda_{n+1} = 1, \lambda_i \geqslant 0, 1 \leqslant i \leqslant n+1),$$

where $\mathfrak{P}_i$ *is* $\mathbf{a}_i . \mathbf{x} = 1$.

This follows by duality in R^n from Theorem 18, since the dual of an intersection of closed half-spaces is the convex cover of the points dual to the bounding hyperplanes of these half-spaces.

4. Kirchberger's theorem

In this paragraph an example is given of the use of Helly's theorem to prove a result due initially to Kirchberger.

THEOREM 21. *Let* $\mathfrak{X}_1$ *and* $\mathfrak{X}_2$ *be two finite sets of points in* R^n. *If, given any subset* $\mathfrak{Y}$ *consisting of* $n+2$ *points selected from the union* $\mathfrak{X}_1 \cup \mathfrak{X}_2$, *it is possible to find a hyperplane that strictly separates* $\mathfrak{Y} \cap \mathfrak{X}_1$ *from* $\mathfrak{Y} \cap \mathfrak{X}_2$, *then there is a hyperplane that strictly separates* $\mathfrak{X}_1$ *from* $\mathfrak{X}_2$.

Let $\mathbf{x}$ be a point of R^n and in the dual space $\mathsf{E}(n)$ of dimension $n+1$ with variables $(a_1, a_2, \ldots, a_n, b)$ define two open half-spaces $\mathfrak{J}(\mathbf{x})$ and $\mathfrak{W}(\mathbf{x})$ by

$$(a_1, \ldots, a_n, b) \in \mathfrak{J}(\mathbf{x}) \quad \text{if} \quad a_1 x_1 + \ldots + a_n x_n + b > 0, \quad (2\cdot4\cdot1)$$

$$(a_1, \ldots, a_n, b) \in \mathfrak{W}(\mathbf{x}) \quad \text{if} \quad a_1 x_1 + \ldots + a_n x_n + b < 0. \quad (2\cdot4\cdot2)$$

Consider the class of half-spaces $\mathfrak{J}(\mathbf{x}), \mathbf{x} \in \mathfrak{X}_1$ and $\mathfrak{W}(\mathbf{x}), \mathbf{x} \in \mathfrak{X}_2$. The hypotheses of the theorem ensure that every $n+2$ of these half-spaces have a point in common and therefore by Helly's theorem there is a point common to them all. This point is, say, $(a_1', \ldots, a_n', b')$, and the hyperplane $a_1' x_1 + \ldots + a_n' x_n + b' = 0$ strictly separates $\mathfrak{X}_1$ from $\mathfrak{X}_2$.

The number $n+2$ used in Theorem 21 cannot be reduced. Consider, for example, the case in which $\mathfrak{X}_1$ is the vertices of a regular simplex and $\mathfrak{X}_2$ is the centroid of the simplex. Every subset $\mathfrak{Y}$ consisting of $n+1$ points of $\mathfrak{X}_1 \cup \mathfrak{X}_2$ is such that $\mathfrak{Y} \cap \mathfrak{X}_1$ can be separated from $\mathfrak{Y} \cap \mathfrak{X}_2$. But $\mathfrak{X}_1$ cannot be separated from $\mathfrak{X}_2$.

5. Horn's extensions of Helly's theorem

An extension of Helly's theorem has been given by A. Horn, in which the idea is to relax the intersection condition on the class of sets in R^n. Whereas the original Helly theorem requires that every subclass with $n+1$ members has a common point, the modified form only requires that every subclass of k members, $1 \leqslant k \leqslant n$, has a common point. The conclusion is, of course, weaker than that of Helly's theorem.

THEOREM 22. *A finite collection of bounded closed convex sets in R^n has the property that every k of the sets have a point in common, $1 \leqslant k \leqslant n$. Then, given any $n-k$-dimensional linear space $\mathfrak{L}_1$, there can be found an $n-k+1$-dimensional linear space $\mathfrak{L}_2$ such that $\mathfrak{L}_2 \supset \mathfrak{L}_1$ and $\mathfrak{L}_2$ intersects each of the convex sets of the given collection.*

Consider first the case $k=n$. We have to show that if every n-membered subclass of a given class of convex sets has a common point, then through any given point there passes a line which meets each of the sets.

Suppose that the class contains N sets. We shall prove the result by induction on N. The result is trivial if $N=n$. Suppose inductively that the theorem is true of every class of $N-1$ sets in R^n and let the given class of sets be $\mathfrak{X}_1, \mathfrak{X}_2, \ldots, \mathfrak{X}_N$. Let the given point be $\mathbf{y}$. We assume that $\mathbf{y}$ does not belong to any of the sets $\mathfrak{X}_1, \ldots, \mathfrak{X}_N$, since if it did the required result would follow immediately from the induction hypothesis applied to the remaining $N-1$ of the $\mathfrak{X}_i$.

Let $\mathfrak{S}$ be the surface of an n-dimensional sphere whose centre is $\mathbf{y}$. Denote the intersection $\mathfrak{C}(\mathbf{y}, \mathfrak{X}_i) \cap \mathfrak{S}$ by $\mathfrak{X}'_i$. Each set $\mathfrak{X}'_i$ is contained in an open hemisphere on $\mathfrak{S}$. By the induction hypothesis we can find a point, say $\mathbf{s}_j$, of the intersection $\bigcap_{i \neq j} \mathfrak{X}'_i$, $j=1, 2, \ldots, N$. Either $\mathbf{s}_j$ and $\mathfrak{X}'_j$ lie in an open hemisphere on $\mathfrak{S}$ or the line $\mathbf{y}\mathbf{s}_j$ meets $\mathfrak{X}'_j$. In this second case the theorem is proved. In the first case there is a hyperplane $\mathfrak{P}$ through $\mathbf{y}$ such that $\mathbf{s}_j$ and $\mathfrak{X}'_j$ lie on the same side of $\mathfrak{P}$. Since $\mathbf{s}_i \in \mathfrak{X}'_j$, $i \neq j$, it follows that all the points $\mathbf{s}_i$, $i=1, 2, \ldots, N$, lie on the same side of $\mathfrak{P}$. On the line $\mathbf{y}\mathbf{s}_i$ take a point $\mathbf{x}_j^{(i)}$ of $\mathfrak{X}_j$, $j \neq i$. Form the convex cover of the set $\bigcup_{i \neq j} \mathbf{x}_j^{(i)}$ and denote it by $\mathfrak{Y}_j$, $j=1, 2, \ldots, N$. Then $\mathfrak{Y}_j \subset \mathfrak{X}_j$ and if we project the sets $\mathfrak{Y}_j$ from $\mathbf{y}$ onto a hyperplane parallel to $\mathfrak{P}$, the sets $\mathfrak{Y}_j$ become a class of bounded closed convex sets of which every $N-1$ intersect. Now $N \geqslant n+1$, and it follows from Helly's theorem in R^{n-1} that there is a point common to all these projections. The line joining $\mathbf{y}$ to this point intersects all the sets $\mathfrak{X}_j$.

Thus the case $k=n$ is proved.

Next suppose that $1 \leqslant k < n$. Let the given $(n-k)$-dimensional linear manifold be $\mathfrak{Q}$ and a k-dimensional linear manifold perpendicular to $\mathfrak{Q}$ be $\mathfrak{R}$. Project the whole space R^n onto $\mathfrak{R}$ by means of $(n-k)$-dimensional linear manifolds parallel to $\mathfrak{Q}$. Then in $\mathfrak{R}$ we have a class of convex sets every k of which have a common point, and $\mathfrak{R}$ is essentially an R^k. $\mathfrak{Q}$ projects into a point $\mathbf{q}$, and by the part of the theorem already proved there is a line through $\mathbf{q}$ in $\mathfrak{R}$ which meets the projection of each convex set of the given class. This line is the projection of an $(n-k+1)$-dimensional linear manifold that contains $\mathfrak{Q}$ and intersects every member of the original class of convex sets.

Thus the theorem is proved.

Chapter 3

GENERAL PROPERTIES OF CONVEX FUNCTIONS

In this chapter we develop the properties of convex functions. For technical reasons it is desirable to have available a variety of conditions for a function to be convex; sometimes it is more convenient to apply one condition than another. We give such conditions in §§ 3 and 4. It should be noted that the condition given in § 3 is a local condition in the sense that it is given in terms of derivatives. The condition has to be satisfied at all points of the domain of definition of the function and when it is satisfied we conclude that the defining inequality of § 1 holds. This inequality may be regarded as the global property obtained from the condition of § 3 by a process of 'integration'.

There are several ways in which convex sets are related to convex functions, and we shall discuss two of them in § 5.

1. The definition of a convex function: boundedness and continuity

We shall consider convex functions $f(\mathbf{x})$ defined for points $\mathbf{x} = (x_1, \ldots, x_n)$ which belong to a convex subset $\mathfrak{X}$ of R^n.

Definition. *The function* $f(\mathbf{x})$ *is said to be convex if*

$$f(\lambda\mathbf{x}_1 + \mu\mathbf{x}_2) \leqslant \lambda f(\mathbf{x}_1) + \mu f(\mathbf{x}_2) \quad (\lambda \geqslant 0, \mu \geqslant 0, \lambda + \mu = 1), \quad (3\cdot1\cdot1)$$

where $\mathbf{x}_1$ *and* $\mathbf{x}_2$ *are any two points of* $\mathfrak{X}$.

If the function $-f(\mathbf{x})$ is convex then we say that $f(\mathbf{x})$ is *concave*. If $f(\mathbf{x})$ is both convex and concave then it is a linear function of the variables $x_1, \ldots, x_n$.

Theorem 23. *A convex function* $f(\mathbf{x})$ *defined over a convex set* $\mathfrak{X}$ *is bounded above on any closed and bounded set* $\mathfrak{X}_1$ *contained in the relative interior of* $\mathfrak{X}$.

By the Heine-Borel theorem $\mathfrak{X}_1$ can be covered by a finite number of simplexes whose vertices belong to $\mathfrak{X}$. Thus it is sufficient to show that $f(\mathbf{x})$ is bounded above on every simplex contained in $\mathfrak{X}$. Consider the r-dimensional simplexes contained in $\mathfrak{X}$. By the definition (3·1·1) $f(\mathbf{x})$ is bounded above on any one-dimensional simplex contained in $\mathfrak{X}$. Assume inductively that $f(\mathbf{x})$ is bounded above on any $(r-1)$-dimensional simplex contained in $\mathfrak{X}$ and consider the r-dimensional simplex $\mathfrak{K}$ whose vertices $\mathbf{x}_1, \ldots, \mathbf{x}_{r+1}$ belong to $\mathfrak{X}$. If $\mathbf{x} \in \mathfrak{K}$ and $\mathbf{x} \neq \mathbf{x}_{r+1}$, then

$$\mathbf{x} = \lambda_1 \mathbf{x}_1 + \lambda_2 \mathbf{x}_2 + \ldots + \lambda_{r+1} \mathbf{x}_{r+1},$$

where

$$\lambda_1 + \ldots + \lambda_{r+1} = 1 \quad (\lambda_i \geqslant 0;\ i = 1, 2, \ldots, r+1;\ \lambda_{r+1} < 1),$$

and this may be written as

$$\mathbf{x} = \mu \mathbf{z} + \lambda_{r+1} \mathbf{x}_{r+1}, \tag{3·1·2}$$

where

$$\mu = 1 - \lambda_{r+1} = \lambda_1 + \ldots + \lambda_r, \quad \mathbf{z} = (\lambda_1 \mathbf{x}_1 + \ldots + \lambda_r \mathbf{x}_r)/(\lambda_1 + \ldots + \lambda_r).$$

Now $\mathbf{z}$ is a point of an $(r-1)$-dimensional simplex in $\mathfrak{X}$. Thus $f(\mathbf{z}) \leqslant M$, where M depends on $\mathbf{x}_1, \ldots, \mathbf{x}_r$ but not upon $\lambda_1, \ldots, \lambda_r$. By (3·1·1) and (3·1·2) we have

$$f(\mathbf{x}) \leqslant \max(M, f(\mathbf{x}_{r+1})),$$

and the next stage of the induction has been established.

This completes the proof of the theorem.

Theorem 24. *A convex function $f(\mathbf{x})$ defined over a convex set $\mathfrak{X}$ is continuous at every point of the relative interior of $\mathfrak{X}$.*

Changing the notation, we obtain from (3·1·1)

$$f(\mathbf{x}_1) - f(\mathbf{x}_2) \leqslant \lambda[f(\mathbf{x}_2 + (\mathbf{x}_1 - \mathbf{x}_2)/\lambda) - f(\mathbf{x}_2)] \quad (0 < \lambda \leqslant 1). \tag{3·1·3}$$

Replace $\mathbf{x}_1$ by $\mathbf{x}_2 + \lambda\mathbf{y}$, then (3·1·3) becomes

$$f(\mathbf{x}_2 + \lambda\mathbf{y}) - f(\mathbf{x}_2) \leqslant \lambda[f(\mathbf{x}_2 + \mathbf{y}) - f(\mathbf{x}_2)].$$

Since, by Theorem 23, $f(\mathbf{x})$ is bounded in some neighbourhood of $\mathbf{x}_2$, it follows that for λ sufficiently small, $0 < \lambda \leqslant 1$,

$$f(\mathbf{x}_2 + \lambda\mathbf{y}) - f(\mathbf{x}_2) \leqslant \lambda M. \tag{3·1·4}$$

In (3·1·4) replace $\mathbf{x}_2 + \lambda\mathbf{y}$ by $\mathbf{x}_2$, $\mathbf{x}_2$ by $\mathbf{x}_2 - \lambda\mathbf{y}$ and then change $\mathbf{y}$ to $-\mathbf{y}$; we obtain

$$f(\mathbf{x}_2) - f(\mathbf{x}_2 + \lambda\mathbf{y}) \leqslant \lambda M. \tag{3·1·5}$$

The inequalities (3·1·4) and (3·1·5) imply that $f(\mathbf{x})$ is continuous at $\mathbf{x}_2$. Thus the theorem is proved.

THEOREM 25. *If $f(\mathbf{x})$ is convex and $f(\mathbf{O})=0$, then $f(\mu\mathbf{x})/\mu$ is an increasing function of μ for $\mu>0$.*

We have to show that

$$\mu_2 f(\mu_1\mathbf{x}) \geqslant \mu_1 f(\mu_2\mathbf{x}), \tag{3·1·6}$$

where $\mu_1 \geqslant \mu_2 > 0$. But since $f(\mathbf{O})=0$ this is the same as

$$f(\mu_2\mathbf{x}) \leqslant \frac{\mu_2}{\mu_1} f(\mu_1\mathbf{x}) + \frac{\mu_1-\mu_2}{\mu_1} f(\mathbf{O}),$$

and this last inequality follows from the convexity of $f(\mathbf{x})$ and the definition (3·1·1).

COROLLARY. *If $f(\mathbf{x})$ is convex, then $(f(\mathbf{x}_0+\mu\mathbf{x})-f(\mathbf{x}_0))/\mu$, is an increasing function of μ for $\mu>0$.*

This follows from the theorem applied to the function $f(\mathbf{x}_0+\mathbf{x})-f(\mathbf{x}_0)$.

An important type of convex function is the gauge function whose definition is as follows:

DEFINITION. *A gauge function is a convex function which is positively homogeneous, i.e. $f(\mu\mathbf{x})=\mu f(\mathbf{x})$ for all $\mu \geqslant 0$.*

Alternatively a gauge function is a function defined over a convex cone $\mathfrak{X}$ for which

(i) $f(\mu\mathbf{x})=\mu f(\mathbf{x}) \quad (\mu \geqslant 0, \mathbf{x}\in\mathfrak{X}, \mu\mathbf{x}\in\mathfrak{X})$,

(ii) $f(\mathbf{x}+\mathbf{y}) \leqslant f(\mathbf{x})+f(\mathbf{y}) \quad (\mathbf{x},\mathbf{y},\mathbf{x}+\mathbf{y}\in\mathfrak{X})$.

It is easy to see that these two definitions are equivalent. A gauge function is not necessarily positive.

DEFINITION. *A vector $\mathbf{x}$ is said to be a linearity vector of the gauge function $f(\mathbf{x})$ if*

$$f(\mu\mathbf{x})=\mu f(\mathbf{x}) \quad \textit{all real } \mu.$$

THEOREM 26. *If $\mathbf{x}_1$ and $\mathbf{x}_2$ are linearity vectors of a gauge function $g(\mathbf{x})$ then so is $\lambda\mathbf{x}_1+\mu\mathbf{x}_2$, where λ, μ are any pair of real numbers.*

By the definition of a gauge function we have

$$g(\mathbf{x}_1+\mathbf{x}_2) \leqslant g(\mathbf{x}_1)+g(\mathbf{x}_2), \tag{3·1·7}$$

and as $\mathbf{x}_1$ and $\mathbf{x}_2$ are linearity vectors,

$$g(\mathbf{x}_1) = -g(-\mathbf{x}_1), \quad g(\mathbf{x}_2) = -g(-\mathbf{x}_2). \tag{3·1·8}$$

By (3·1·7) with $\mathbf{x}_1, \mathbf{x}_2$ replaced by $-\mathbf{x}_1$, $-\mathbf{x}_2$, we have

$$g(-\mathbf{x}_1 - \mathbf{x}_2) \leqslant g(-\mathbf{x}_1) + g(-\mathbf{x}_2).$$

Hence (3·1·8) gives

$$-g(-\mathbf{x}_1 - \mathbf{x}_2) \geqslant g(\mathbf{x}_1) + g(\mathbf{x}_2). \tag{3·1·9}$$

From (3·1·9) and (3·1·7) we have

$$g(\mathbf{x}_1 + \mathbf{x}_2) \leqslant -g(-\mathbf{x}_1 - \mathbf{x}_2). \tag{3·1·10}$$

But $g(\mathbf{x})$ is a gauge function and thus

$$g(\mathbf{x}_1 + \mathbf{x}_2) + g(-\mathbf{x}_1 - \mathbf{x}_2) \geqslant g(\mathbf{O}) = 0. \tag{3·1·11}$$

From (3·1·10) and (3·1·11) we see that $g(\mathbf{x}_1 + \mathbf{x}_2)$ is equal to $-g(-\mathbf{x}_1 - \mathbf{x}_2)$, and this, combined with the positive homogeneity of a gauge function implies that $\lambda\mathbf{x}_1 + \mu\mathbf{x}_2$ is a linearity vector of $g(\mathbf{x})$.

Thus the theorem is proved.

THEOREM 27. *Suppose that* $\mathfrak{J}$ *is an index class and that for* $i \in \mathfrak{J}$ $f_i(\mathbf{x})$ *is a gauge function defined over a convex set* $\mathfrak{X}$. *If this family of functions is uniformly bounded above, the function* $g(\mathbf{x})$ *defined by* $g(\mathbf{x}) = \sup_{i \in \mathfrak{J}} f_i(\mathbf{x})$ *is a gauge function defined over* $\mathfrak{X}$.

We have to show that

(i) $g(\mu\mathbf{x}) = \mu g(\mathbf{x}) \quad (\mu \geqslant 0)$,

(ii) $g(\mathbf{x} + \mathbf{y}) \leqslant g(\mathbf{x}) + g(\mathbf{y})$.

Now $\quad g(\mu\mathbf{x}) = \sup_{i \in \mathfrak{J}} f_i(\mu\mathbf{x}) = \sup_{i \in \mathfrak{J}} \mu f_i(\mathbf{x}) = \mu \sup_{i \in \mathfrak{J}} f_i(\mathbf{x}) = \mu g(\mathbf{x})$,

and $\quad g(\mathbf{x}) + g(\mathbf{y}) = \sup_{i \in \mathfrak{J}} f_i(\mathbf{x}) + \sup_{i \in \mathfrak{J}} f_i(\mathbf{y}) \geqslant \sup_{i \in \mathfrak{J}} (f_i(\mathbf{x}) + f_i(\mathbf{y}))$

$$\geqslant \sup_{i \in \mathfrak{J}} f_i(\mathbf{x} + \mathbf{y}) = g(\mathbf{x} + \mathbf{y}).$$

Thus the theorem is proved.

EXERCISES 3·1

1. Show that if $f(\mathbf{x})$ is a continuous function defined over a convex set, and if

$$f(\tfrac{1}{2}\mathbf{x}_1 + \tfrac{1}{2}\mathbf{x}_2) \leqslant \tfrac{1}{2}f(\mathbf{x}_1) + \tfrac{1}{2}f(\mathbf{x}_2)$$

for every pair of points of the convex set, then $f(\mathbf{x})$ is a convex function.

2. Show that if $f(x)$ is a convex differentiable function defined over the interval $a \leqslant x \leqslant b$ of the real axis, then $f'(x)$ is an increasing function of x, and conversely.

3. $f(x)$ is a convex differentiable function of the real variable x defined over $a \leqslant x \leqslant b$ and $\phi(t)$ is defined by

$$\phi(t) = \sup_{a \leqslant x \leqslant b} (xt - f(x)).$$

Show that $\phi(t)$ is a convex function of t for all real values of t.

2. The directional derivatives of a convex function

There is an important connexion between a given convex function $f(\mathbf{x})$ and a certain gauge function, which may be summarized by the statement that $f(\mathbf{x})$ behaves, in the neighbourhood of a point $\mathbf{x}_0$, as though $f(\mathbf{x}) - f(\mathbf{x}_0)$ were approximately a gauge function of $\mathbf{x} - \mathbf{x}_0$. The corresponding statement concerning convex sets is that near a frontier point $\mathbf{x}_0$ the convex set approximates to a certain convex cone whose vertex is $\mathbf{x}_0$.

Write
$$\delta f(\mathbf{x}_0; \mathbf{x}) = \inf_{\mu > 0} (f(\mathbf{x}_0 + \mu \mathbf{x}) - f(\mathbf{x}_0))/\mu. \qquad (3\cdot2\cdot1)$$

The function $\delta f(\mathbf{x}_0; \mathbf{x})$ is said to be the *directional derivative*† of $f(\mathbf{x})$ at $\mathbf{x}_0$ in the direction of $\mathbf{x}$. It should be noted that $\mathbf{x}$ is not a direction vector; it has magnitude as well as direction. The directional derivative is a function of $2n$ variables of which n are the coordinates of $\mathbf{x}$ and n the coordinates of $\mathbf{x}_0$. By the corollary to Theorem 25 we have the equality

$$\delta f(\mathbf{x}_0; \mathbf{x}) = \lim_{\mu \to 0+} (f(\mathbf{x}_0 + \mu \mathbf{x}) - f(\mathbf{x}_0))/\mu, \qquad (3\cdot2\cdot2)$$

and this relation may be taken as the definition.

The function $\delta f(\mathbf{x}_0; \mathbf{x})$ is a gauge function of $\mathbf{x}$. For by (3·2·2) when $\lambda \geqslant 0$ we have

$$\delta f(\mathbf{x}_0; \lambda \mathbf{x}) = \lambda \delta f(\mathbf{x}_0; \mathbf{x}), \qquad (3\cdot2\cdot3)$$

and further from (3·1·1) on writing $\mathbf{x}_1 = \mathbf{x}_0 + 2h\mathbf{y}$, $\mathbf{x}_2 = \mathbf{x}_0 + 2h\mathbf{z}$, where $h \geqslant 0$, and taking the particular values $\lambda = \mu = \frac{1}{2}$, we have

$$f(\mathbf{x}_0 + h(\mathbf{y} + \mathbf{z})) \leqslant \tfrac{1}{2} f(\mathbf{x}_0 + 2h\mathbf{y}) + \tfrac{1}{2} f(\mathbf{x}_0 + 2h\mathbf{z}). \qquad (3\cdot2\cdot4)$$

† It is a one-sided Gateaux differential.

Subtract $f(\mathbf{x}_0)$ from each side, divide by h and let h tend to zero. Then

$$\delta f(\mathbf{x}_0;\ \mathbf{y}+\mathbf{z}) \leqslant \delta f(\mathbf{x}_0;\ \mathbf{y}) + \delta f(\mathbf{x}_0;\ \mathbf{z}). \tag{3·2·5}$$

The relations (3·2·3) and (3·2·5) show that $\delta f(\mathbf{x}_0;\ \mathbf{x})$ is a gauge function of $\mathbf{x}$.

If we put $\mu = 1$ in (3·2·1) we obtain the important inequality

$$\delta f(\mathbf{x}_0;\ \mathbf{x}) \leqslant f(\mathbf{x}_0+\mathbf{x}) - f(\mathbf{x}_0).$$

In particular, if $f(\mathbf{x})$ is a gauge function, we have

$$\delta f(\mathbf{x}_0;\ \mathbf{x}) \leqslant f(\mathbf{x}). \tag{3·2·6}$$

Just as we obtained the directional derivative $\delta f(\mathbf{x}_0;\ \mathbf{x})$ of the given convex function $f(\mathbf{x})$, so we can obtain the directional derivative of $\delta f(\mathbf{x}_0;\ \mathbf{x})$ as a function of $\mathbf{x}$ at $\mathbf{x}_1$ in the direction $\mathbf{x}_2$ and denote this by $\delta^2 f(\mathbf{x}_0;\ \mathbf{x}_1;\ \mathbf{x}_2)$. Since $\delta f(\mathbf{x}_0;\ \mathbf{x})$ is not only a convex function of $\mathbf{x}$ but satisfies the stronger condition of being a gauge function, there are relations between $\delta f(\mathbf{x}_0;\ \mathbf{x})$ and $\delta^2 f(\mathbf{x}_0;\ \mathbf{x}_1;\ \mathbf{x})$ which do not necessarily hold between $f(\mathbf{x})$ and $\delta f(\mathbf{x}_0;\ \mathbf{x})$.

For example

$$\delta^2 f(\mathbf{x}_0;\mathbf{x}_1;\mathbf{x}_1) = \lim_{h\to 0+} \frac{(1+h)\,\delta f(\mathbf{x}_0;\mathbf{x}_1) - \delta f(\mathbf{x}_0;\mathbf{x}_1)}{h} = \delta f(\mathbf{x}_0;\ \mathbf{x}_1).$$

Similarly

$$\delta^2 f(\mathbf{x}_0;\ \mathbf{x}_1;\ -\mathbf{x}_1) = -\delta f(\mathbf{x}_0;\ \mathbf{x}_1). \tag{3·2·7}$$

Thus $\mathbf{x}_1$ is always a linearity vector of $\delta^2 f(\mathbf{x}_0;\ \mathbf{x}_1;\ \mathbf{x})$. Further, if $\mathbf{x}_2$ is a linearity vector of $\delta f(\mathbf{x}_0;\ \mathbf{x})$ then it is also one of $\delta^2 f(\mathbf{x}_0;\ \mathbf{x}_1;\ \mathbf{x})$. For we have to show that if

$$\delta f(\mathbf{x}_0;\ -\mathbf{x}_2) = -\delta f(\mathbf{x}_0;\ \mathbf{x}_2), \tag{3·2·8}$$

then

$$\delta^2 f(\mathbf{x}_0;\ \mathbf{x}_1;\ -\mathbf{x}_2) = -\delta^2 f(\mathbf{x}_0;\ \mathbf{x}_1;\ \mathbf{x}_2). \tag{3·2·9}$$

Now

$$\delta^2 f(\mathbf{x}_0;\ \mathbf{x}_1;\ -\mathbf{x}_2) = \lim_{h\to 0+} \frac{\delta f(\mathbf{x}_0;\mathbf{x}_1 - h\mathbf{x}_2) - \delta f(\mathbf{x}_0;\mathbf{x}_1)}{h}$$

$$\leqslant \lim_{h\to 0+} \frac{\delta f(\mathbf{x}_0;\ -h\mathbf{x}_2)}{h}.$$

Since $\delta f(\mathbf{x}_0;\ \mathbf{x})$ is a gauge function of $\mathbf{x}$ for which (3·2·8) holds, it follows that

$$\delta^2 f(\mathbf{x}_0;\ \mathbf{x}_1;\ -\mathbf{x}_2) \leqslant -\delta f(\mathbf{x}_0;\ \mathbf{x}_2).$$

Similarly $$\delta^2 f(\mathbf{x}_0;\, \mathbf{x}_1;\, \mathbf{x}_2) \leqslant \delta f(\mathbf{x}_0;\, \mathbf{x}_2).$$

Thus $$-\delta^2 f(\mathbf{x}_0;\, \mathbf{x}_1;\, -\mathbf{x}_2) \geqslant \delta^2 f(\mathbf{x}_0;\, \mathbf{x}_1;\, \mathbf{x}_2). \tag{3·2·10}$$

But $\delta^2 f(\mathbf{x}_0;\, \mathbf{x}_1;\, \mathbf{x})$ is a gauge function of $\mathbf{x}$. Thus we have

$$\delta^2 f(\mathbf{x}_0;\, \mathbf{x}_1;\, \mathbf{x}_2) + \delta^2 f(\mathbf{x}_0;\, \mathbf{x}_1;\, -\mathbf{x}_2) \geqslant 0. \tag{3·2·11}$$

Thus (3·2·9) follows from (3·2·10) and (3·2·11).

EXERCISES 3·2

1. Show that the function $f(\mathbf{x}) = x_1^r + \ldots + x_n^r$ is a convex function of $\mathbf{x} = (x_1, \ldots, x_n)$ if $r \geqslant 1$ and $x_i > 0$, $i = 1, 2, \ldots, n$. Find its directional derivative.

2. The convex function $f(\mathbf{x})$, where $\mathbf{x} = (x_1, x_2)$, has continuous partial derivatives of the first order with respect to each variable x_1, x_2. Show that $\delta f(\mathbf{x};\, \mathbf{y})$ is a linear function of $\mathbf{y}$, and that $f(\mathbf{x})$ is a differentiable function of $\mathbf{x}$.

3. $f(\mathbf{x})$ is a given convex function defined for all points $\mathbf{x}$ of a convex set $\mathfrak{X}$. G is the class of gauge functions $g(\mathbf{x})$ such that

$$f(\mathbf{x}) - f(\mathbf{x}_0) \geqslant g(\mathbf{x} - \mathbf{x}_0)$$

and $h(\mathbf{x})$ is defined by $$h(\mathbf{x}) = \sup_{g \in G} g(\mathbf{x}).$$

Show that $$h(\mathbf{x}) = \delta f(\mathbf{x}_0;\, \mathbf{x}).$$

3. Differential conditions for convexity

If $f(\mathbf{x}) = f(x_1, \ldots, x_n)$ has continuous partial derivatives of the second order, then the necessary and sufficient condition that it should be convex at every point $\mathbf{x}$ of an open convex set $\mathfrak{X}$ is that the quadratic form in $\mathbf{t}$

$$Q(\mathbf{x};\, \mathbf{t}) = \sum_{i=1}^{n} \sum_{j=1}^{n} \frac{\partial^2 f}{\partial x_i \partial x_j} t_i t_j \tag{3·3·1}$$

should be non-negative for all real numbers t_i and all $\mathbf{x}$ of $\mathfrak{X}$. Here the coefficients $\dfrac{\partial^2 f}{\partial x_i \partial x_j}$ are all calculated at the point $\mathbf{x}$ concerned.

For we have, when $0<\theta<1$,

$$(1-\theta)f(\mathbf{x})+\theta f(\mathbf{y})-f[(1-\theta)\mathbf{x}+\theta\mathbf{y}]$$
$$=(1-\theta)[f(\mathbf{x})-f((1-\theta)\mathbf{x}+\theta\mathbf{y})]+\theta[f(\mathbf{y})-f((1-\theta)\mathbf{x}+\theta\mathbf{y})]$$
$$=(1-\theta)\left[\sum_{i=1}^{n}\theta(x_i-y_i)\frac{\partial f}{\partial x_i}+\frac{1}{2}\sum_{i,j=1}^{n}{}'\,\theta^2(x_i-y_i)(x_j-y_j)\frac{\partial^2 f}{\partial x_i\partial x_j}\right]$$
$$+\theta\left[\sum_{i=1}^{n}(1-\theta)(y_i-x_i)\frac{\partial f}{\partial x_i}\right.$$
$$\left.+\frac{1}{2}\sum_{i,j=1}^{n}{}''\,(1-\theta)^2(y_i-x_i)(y_j-x_j)\frac{\partial^2 f}{\partial x_i\partial x_j}\right] \quad (3\cdot 3\cdot 2)$$

by the mean-value theorem, where the first-order derivatives are calculated at $(1-\theta)\mathbf{x}+\theta\mathbf{y}$ and, for the sum Σ', the second-order derivatives are calculated at a suitable point on the segment joining $\mathbf{x}$ to $(1-\theta)\mathbf{x}+\theta\mathbf{y}$, whilst for the sum Σ'' they are calculated at a point on the segment joining $\mathbf{y}$ to $(1-\theta)\mathbf{x}+\theta\mathbf{y}$.

The terms in (3·3·2) which involve first-order derivatives cancel, and if $Q(\mathbf{x};\,\mathbf{t})$ is positive semi-definite it follows that

$$(1-\theta)f(\mathbf{x})+\theta f(\mathbf{y})-f[(1-\theta)\mathbf{x}+\theta\mathbf{y}]\geqslant 0,$$

i.e. $f(\mathbf{x})$ is a convex function of $\mathbf{x}$.

On the other hand, if $Q(\mathbf{x};\,\mathbf{t})$ is negative for some $\mathbf{x}$ and $\mathbf{t}$, then there is a small sphere $\mathfrak{S}(\mathbf{x}',r)$ such that $Q(\mathbf{x},\mathbf{t})$ is negative for all $\mathbf{x}\in\mathfrak{S}(\mathbf{x}',r)$. Let $\mathbf{y}\in\mathfrak{S}(\mathbf{x}';r)$ be such that $\mathbf{y}-\mathbf{x}=\lambda\mathbf{t}$. Then the expression (3·3·2) is negative and hence $f(\mathbf{x})$ is not convex.

Exercises 3·3

1. $f(\mathbf{x})$ is a convex function of $\mathbf{x}=(x_1,\ldots,x_n)$ and $g(y)$ is a convex increasing function of y, where y is a real variable and the domain of g contains the range of f. Show that $g(f(\mathbf{x}))$ is a convex function of $\mathbf{x}$.

2. $f_i(x), i=1,2,\ldots,n$, are n convex functions of the real variable $x, a\leqslant x\leqslant b$, and the function of x, $F(x)$, is defined by $F(x)=g(f_1(x),\ldots,f_n(x))$, where g is a convex function of n variables that is increasing in each variable separately. The domain of definition of g is such that it includes the Cartesian product of the ranges of the $f_i(x)$. Show that $F(x)$ is a convex function of x, $a\leqslant x\leqslant b$.

3. For every value of t, $a\leqslant t\leqslant b$, $f(\mathbf{x},t)$ is a convex function of

$\mathbf{x} = (x_1, \ldots, x_n)$ with second-order partial derivatives with respect to the x_i continuous in $\mathbf{x}$ and t. Show that

$$g(\mathbf{x}) = \int_a^b f(\mathbf{x}, t)\,dt$$

is convex.

4. Planar convexity in terms of polar coordinates

The frontier of a plane two-dimensional convex set is a curve with the property that any straight line meets it in at most two points. Conversely, a simple closed Jordan curve with the property that any line meets it in at most two points bounds a convex domain. We can use this property to obtain a condition for convexity in terms of polar coordinates. The n-dimensional analogue of this condition is not given here as it does not lead to convexity but to a condition that a curve in R^n should meet each hyperplane in R^n at most n times.

THEOREM 28. *Let $F(x_1, x_2)$, $G(x_1, x_2)$ be two positive gauge functions, then*

$$\begin{vmatrix} F(x_1,x_2) & x_1 & x_2 \\ F(y_1,y_2) & y_1 & y_2 \\ F(z_1,z_2) & z_1 & z_2 \end{vmatrix} \begin{vmatrix} G(x_1,x_2) & x_1 & x_2 \\ G(y_1,y_2) & y_1 & y_2 \\ G(z_1,z_2) & z_1 & z_2 \end{vmatrix} \geqslant 0 \qquad (3\cdot4\cdot1)$$

for any three points $\mathbf{x} = (x_1, x_2)$, $\mathbf{y} = (y_1, y_2)$, $\mathbf{z} = (z_1, z_2)$.

For if $\mathbf{x}, \mathbf{y}, \mathbf{z}$ are given there exist three real numbers λ, μ, ν not all zero such that

$$\lambda\mathbf{x} + \mu\mathbf{y} + \nu\mathbf{z} = 0, \qquad (3\cdot4\cdot2)$$

and the inequality (3·4·1) will follow if we can show that

$$[\lambda F(\mathbf{x}) + \mu F(\mathbf{y}) + \nu F(\mathbf{z})]\,[\lambda G(\mathbf{x}) + \mu G(\mathbf{y}) + \nu G(\mathbf{z})] \geqslant 0. \qquad (3\cdot4\cdot3)$$

Of the three numbers λ, μ, ν, two at least are either non-positive or non-negative. We may suppose without real loss of generality that μ, ν are non-negative. Either $\lambda \geqslant 0$, in which case (3·4·3) follows from the fact that F and G are positive functions, or $\lambda < 0$, in which case, since F is a gauge function, we have

$$F(\mathbf{x}) = F\left(-\frac{\mu}{\lambda}\mathbf{y} + -\frac{\nu}{\lambda}\mathbf{z}\right) \leqslant F\left(-\frac{\mu}{\lambda}\mathbf{y}\right) + F\left(-\frac{\nu}{\lambda}\mathbf{z}\right)$$

$$= -\frac{\mu}{\lambda}F(\mathbf{y}) - \frac{\nu}{\lambda}F(\mathbf{z}). \qquad (3\cdot4\cdot4)$$

Hence $$\lambda F(\mathbf{x})+\mu F(\mathbf{y})+\nu F(\mathbf{z})\geqslant 0.$$

Since a similar inequality holds for $G(\mathbf{x})$ we have (3·4·3) again.

Suppose now that the origin is an interior point of the convex plane set $\mathfrak{X}$ and that the boundary curve has the equation $r=f(\theta)$ in polar coordinates. Put

$$\mathbf{x}=(\cos\theta,\sin\theta),\quad \mathbf{y}=(\cos\phi,\sin\phi),\quad \mathbf{z}=(\cos\psi,\sin\psi).$$

Let $G(\mathbf{x})=(x_1^2+x_2^2)^{\frac{1}{2}}$, $F(\mathbf{x})=G(\mathbf{x})/f(\theta)$, where θ is the direction from the origin to $\mathbf{x}=(x_1,x_2)$, then we have

$$\begin{vmatrix} [f(\theta)]^{-1} & \cos\theta & \sin\theta \\ [f(\phi)]^{-1} & \cos\phi & \sin\phi \\ [f(\psi)]^{-1} & \cos\psi & \sin\psi \end{vmatrix} \bigg/ \begin{vmatrix} 1 & \cos\theta & \sin\theta \\ 1 & \cos\phi & \sin\phi \\ 1 & \cos\psi & \sin\psi \end{vmatrix} \geqslant 0, \qquad (3\cdot4\cdot5)$$

and this inequality is the condition for convexity in terms of polar coordinates.

Exercises 3·4

1. Prove that if $r=f(\theta)$ is defined, positive and periodic with period 2π and further satisfies (3·4·5), then the set of points $\mathbf{x}=(x_1,x_2)$ for which $x_1^2+x_2^2\leqslant[f(\theta)]^2$, where θ is the direction from the origin to (x_1,x_2), is convex.

5. The distance and support functions of a convex set

Let $\mathfrak{X}$ be a closed bounded n-dimensional convex set containing the origin as an interior point and let $\mathfrak{X}^*$ be the dual convex set.

Definition. *The distance function of* $\mathfrak{X}$, $F(\mathbf{x})$, *is defined by*

$$F(\mathbf{x})=\inf\{\rho:\rho>0,\mathbf{x}\in\rho\mathfrak{X}\}. \qquad (3\cdot5\cdot1)$$

The support function of $\mathfrak{X}$, $H(\mathbf{l})$, *is defined by*

$$H(\mathbf{l})=\inf\{\rho:\rho>0,\mathbf{l}\in\rho\mathfrak{X}^*\}. \qquad (3\cdot5\cdot2)$$

These definitions can be put in other forms which are sometimes more convenient. Since $(\rho\mathfrak{X})^*=\frac{1}{\rho}\mathfrak{X}^*$ we have, for $\mathbf{x}\in\rho\mathfrak{X}$,

$$\sup_{\mathbf{l}\in\mathfrak{X}^*}(\mathbf{l}.\mathbf{x})=\rho\sup_{\frac{1}{\rho}\mathbf{l}\in(\rho\mathfrak{X})^*}\left(\frac{1}{\rho}\mathbf{l}.\mathbf{x}\right)\leqslant\rho$$

by the definition of $(\rho\mathfrak{X})^*$. Hence

$$\sup_{\mathbf{l}\in\mathfrak{X}^*}(\mathbf{l}.\mathbf{x}) \leqslant F(\mathbf{x}).$$

On the other hand, if $\mathbf{x} \notin \rho\mathfrak{X}$, there exists $\frac{1}{\rho}\mathbf{l}$ of $\frac{1}{\rho}(\mathfrak{X})^*$ such that

$$\frac{1}{\rho}\mathbf{l}.\mathbf{x} > 1.$$

Thus $$\sup_{\mathbf{l}\in\mathfrak{X}^*}(\mathbf{l}.\mathbf{x}) > \rho,$$

and since this is true for every $\rho < F(\mathbf{x})$ it follows that

$$\sup_{\mathbf{l}\in\mathfrak{X}^*}(\mathbf{l}.\mathbf{x}) \geqslant F(\mathbf{x}).$$

Thus $$F(\mathbf{x}) = \sup_{\mathbf{l}\in\mathfrak{X}^*}(\mathbf{l}.\mathbf{x}),$$

and similarly $$H(\mathbf{l}) = \sup_{\mathbf{x}\in\mathfrak{X}}(\mathbf{l}.\mathbf{x}).$$

The set $\mathfrak{X}$ is defined by $\mathfrak{X} = \{\mathbf{x}: F(\mathbf{x}) \leqslant 1\}$ and similarly the set $\mathfrak{X}^*$ is the set defined by $\mathfrak{X}^* = \{\mathbf{l}: H(\mathbf{l}) \leqslant 1\}$.

Although $F(\mathbf{x})$ and $H(\mathbf{l})$ have only been defined for particular types of convex set, the function $H(\mathbf{l}) = \sup_{\mathbf{x}\in\mathfrak{X}}(\mathbf{l}.\mathbf{x})$ is quite properly defined for any bounded convex set $\mathfrak{X}$ or indeed for any bounded set.

Since $\mathbf{l}.\mathbf{x}$ is a gauge function of $\mathbf{x}$ for each fixed $\mathbf{l}$ it follows from Theorem 27 that $F(\mathbf{x})$ is a gauge function of $\mathbf{x}$, defined for all $\mathbf{x}$. Similarly $H(\mathbf{l})$ is a gauge function of $\mathbf{l}$ defined for all $\mathbf{l}$.

We have introduced the support and distance functions in such a manner that there is a complete duality between them. $F(\mathbf{x})$ is the distance function of $\mathfrak{X}$ and it is also the support function of $\mathfrak{X}^*$; $H(\mathbf{l})$ is the support function of $\mathfrak{X}$ and the distance function of $\mathfrak{X}^*$; where we use $\mathbf{x}$ as coordinates in R^n and $\mathbf{l}$ as coordinates in the dual space $(\mathsf{R}^n)^*$.

It is not possible to use the distance function when the origin is not an interior point of $\mathfrak{X}$ unless the definition is modified in some way. But the support function can be used whether the origin is an interior point of $\mathfrak{X}$ or not. It is here that the duality breaks down.

It is convenient to have both functions available. For example, if $\mathfrak{X}_1$ and $\mathfrak{X}_2$ are two convex sets which both contain the origin as an interior point, and if $F_1(\mathbf{x})$, $F_2(\mathbf{x})$, $H_1(\mathbf{l})$, $H_2(\mathbf{l})$ are the corresponding distance and support functions, then the distance function of $\mathfrak{X}_1 \cap \mathfrak{X}_2$ is $\max\{F_1(\mathbf{x}), F_2(\mathbf{x})\}$ and the support function of $\mathfrak{H}(\mathfrak{X}_1 \cup \mathfrak{X}_2)$ is $\max\{H_1(\mathbf{l}), H_2(\mathbf{l})\}$, but the support function of $\mathfrak{X}_1 \cap \mathfrak{X}_2$ and the distance function of $\mathfrak{H}(\mathfrak{X}_1 \cup \mathfrak{X}_2)$ are much more complicated. An example where the behaviour of the support function is simpler than that of the distance function is given by considering the effect of moving the origin to the point $\mathbf{a}$. Then $H(\mathbf{l})$ becomes $H_1(\mathbf{l})$ where $H_1(\mathbf{l}) = H(\mathbf{l}) + \mathbf{a}.\mathbf{l}$. The expression for $F_1(\mathbf{x})$ cannot be simply related to that for $F(\mathbf{x})$ since we do not know which $\mathbf{l}$ of $\mathfrak{X}$ is involved in calculating $F_1(\mathbf{x})$; it may be a different vector $\mathbf{l}$ which maximizes the expression $\mathbf{l}(\mathbf{x}-\mathbf{a})$ from that which maximizes $\mathbf{l}.\mathbf{x}$.

Examples of distance and support functions are:

(i) a sphere centre $\mathbf{a}$ and radius r,

$$F(\mathbf{x}) = |\mathbf{x}|^2/(\mathbf{a}.\mathbf{x} + [(r^2 - |\mathbf{a}|^2)|\mathbf{x}|^2 + (\mathbf{a}.\mathbf{x})^2]^{\frac{1}{2}}),$$

$$H(\mathbf{l}) = r|\mathbf{l}| + \mathbf{a}.\mathbf{l},$$

(ii) a cube of side length $2s$ and centre at the origin, with edges parallel to the axes,

$$F(\mathbf{x}) = \max_{1 \leqslant \nu \leqslant n} |x_\nu|/s,$$

$$H(\mathbf{l}) = \sum_{\nu=1}^{n} s|l_\nu|,$$

(iii) an n-dimensional analogue of the regular octahedron with vertices distant s from the origin (i.e. similar to the dual of (ii))

$$F(\mathbf{x}) = \sum_{\nu=1}^{n} |x_\nu|/s,$$

$$H(\mathbf{l}) = \max_{1 \leqslant \nu \leqslant n} s|l_\nu|.$$

The frontier points of a convex set in terms of its support function.

Let $\mathbf{l}'$ be a vector in a fixed direction, then

$$\mathbf{x}.\mathbf{l}' = H(\mathbf{l}')$$

is a support hyperplane to $\mathfrak{X}$ which meets $\mathfrak{X}$ in an at most $(n-1)$-dimensional convex set, say $\mathfrak{Y}$. We shall find the support function of $\mathfrak{Y}$ in terms of the support function of $\mathfrak{X}$. If $\mathbf{y} \in \mathfrak{Y}$, then for any vector $\mathbf{l}$

$$\mathbf{y}.\mathbf{l} \leqslant H(\mathbf{l}), \tag{3·5·3}$$

in particular if $\mathbf{l} = \mathbf{l}' + h\mathbf{m}$, where $h > 0$, then (3·5·3) implies that

$$\mathbf{y}.\mathbf{m} \leqslant \frac{H(\mathbf{l}' + h\mathbf{m}) - H(\mathbf{l}')}{h}. \tag{3·5·4}$$

Let $h \to 0+$ in (3·5·4) and we obtain

$$\mathbf{y}.\mathbf{m} \leqslant \delta H(\mathbf{l}'; \mathbf{m}). \tag{3·5·5}$$

Thus every point of $\mathfrak{Y}$ satisfies (3·5·5) for every vector $\mathbf{m}$. On the other hand, any point $\mathbf{y}$ which satisfies (3·5·5) must also satisfy

$$\mathbf{y}.\mathbf{m} \leqslant H(\mathbf{m}), \tag{3·5·6}$$

since $\delta H(\mathbf{l}'; \mathbf{m}) \leqslant H(\mathbf{m})$ by (3·2·6). Thus if $\mathbf{y}$ satisfies (3·5·5) for all $\mathbf{m}$ we shall have $\mathbf{y} \in \mathfrak{X}$. But further since by (3·2·7) applied to the gauge function $H(\mathbf{l})$ in place of $\delta f(\mathbf{x}_0; \mathbf{x})$

$$\delta H(\mathbf{l}'; \mathbf{l}') = -\delta H(\mathbf{l}'; -\mathbf{l}') = H(\mathbf{l}'),$$

any point $\mathbf{y}$ which satisfies (3·5·5) for all $\mathbf{m}$ also satisfies

$$\mathbf{y}.\mathbf{l}' = H(\mathbf{l}')$$

and must therefore belong to $\mathfrak{Y}$. Thus, finally, the function $\delta H(\mathbf{l}'; \mathbf{m})$ is the support function of the set $\mathfrak{Y}$ (here $\mathbf{l}'$ is a fixed vector and $\mathbf{m}$ is the variable vector).

If the support hyperplane perpendicular to $\mathbf{l}'$ meets $\mathfrak{X}$ in exactly one point, say $\mathbf{y}$, then since the support function of the point $\mathbf{y}$ is $\mathbf{y}.\mathbf{l}$ we have

$$\mathbf{y}.\mathbf{l} = \delta H(\mathbf{l}'; \mathbf{l}).$$

It follows that the partial derivatives of $H(\mathbf{l})$ exist and $\partial H/\partial l_\nu$ is the νth coordinate of $\mathbf{y}$ for $\nu = 1, 2, \ldots, n$.

Every convex set $\mathfrak{X}$ with interior points and with a fixed interior point as origin has associated with it two gauge functions $F(\mathbf{x})$ and $H(\mathbf{l})$. It is also true that given a positive gauge function $G(\mathbf{x})$ there is a convex set $\mathfrak{X}_1$ of which this is the distance function and a convex set $\mathfrak{X}_2$ of which this is the support function. $\mathfrak{X}_1$ is defined to be all $\mathbf{x}$ for which $G(\mathbf{x}) \leqslant 1$ and $\mathfrak{X}_2$ is the dual of $\mathfrak{X}_1$.

If $H(\mathbf{l})$ takes negative values there still exists a convex set of which it is the support function. For let $\mathbf{l}_1, \mathbf{l}_2, \ldots, \mathbf{l}_n$ be n perpendicular directions, then by (3·2·6) we have

$$H(\mathbf{l}) \geqslant \delta H(\mathbf{l}_1;\ \mathbf{l}) \geqslant \delta^2 H(\mathbf{l}_1;\ \mathbf{l}_2;\ \mathbf{l}) \geqslant \ldots \geqslant \delta^n H(\mathbf{l}_1;\ \mathbf{l}_2;\ \ldots;\ \mathbf{l}_n;\ \mathbf{l}).$$

But $\delta^n H(\mathbf{l}_1;\ \mathbf{l}_2;\ \ldots;\ \mathbf{l}_n;\ \mathbf{l})$ has $\mathbf{l}_1, \mathbf{l}_2, \ldots, \mathbf{l}_n$ as linearity vectors, and since these span the whole of R^n the function

$$\delta^n H(\mathbf{l}_1;\ \mathbf{l}_2;\ \ldots;\ \mathbf{l}_n;\ \mathbf{l})$$

is homogeneous in $\mathbf{l}$. Since it is also convex this implies that it is linear, i.e. of the form $\mathbf{a}\,.\,\mathbf{l}$. Thus if we define $G(\mathbf{l})$ by

$$G(\mathbf{l}) = H(\mathbf{l}) - \mathbf{a}\,.\,\mathbf{l},$$

we have

$$G(\mathbf{l}) \geqslant 0 \quad \text{all } \mathbf{l}.$$

Let $\mathfrak{X}_1$ be the set of points $\mathbf{x}$ for which $G(\mathbf{x}) \leqslant 1$. $\mathfrak{X}_1$ may be unbounded but we can still form its dual $\mathfrak{X}_2$. Then $G(\mathbf{l})$ is the support function of $\mathfrak{X}_2$ and $H(\mathbf{l})$ is the support function of a translation of $\mathfrak{X}_2$.

Chapter 4

APPROXIMATIONS TO CONVEX SETS. THE BLASCHKE SELECTION THEOREM

It is essential in many extremal geometrical problems to be able to say whether or not a particular configuration can actually occur. If one knows that the configuration can be attained the problem is usually simplified. This is because one is interested in the extremal values of some functions of the configuration, and those configurations for which this extremal is attained form a class whose structure is essentially simpler than that of the original class. This simplification is particularly striking if the extremal configuration is unique.

The principal object of this chapter is to prove Blaschke's selection theorem. This theorem, which asserts that the class of closed convex subsets of a closed bounded convex set of R^n can be made into a compact metric space, enables one to assert the existence of extremal configurations in many cases. The practical importance of this theorem cannot be overemphasized, and some examples of its use will be given in Chapters 6 and 7.

In this chapter and in all succeeding chapters we shall use the phrase '*convex set*' to mean 'closed bounded convex set', and '*convex body*' to mean 'closed bounded convex set with interior points'. Where a particular convex set is not closed or bounded we shall say so explicitly and we shall sometimes use the phrase 'closed bounded convex set' where it is desirable to emphasize the closed bounded nature of the set. Of course the crucial property throughout this chapter is that of compactness.

1. Classes of convex sets as metric spaces

In the present paragraph we make a number of definitions which will be required later in the proof of Blaschke's theorem.

We consider a closed bounded portion of R^n; for definiteness consider the sphere $\mathfrak{S}(R)$ whose centre is the origin and whose radius is R. Let $\mathfrak{A}$ be the class of closed sets contained in $\mathfrak{S}(R)$. We define a metric in $\mathfrak{A}$ as follows.

Let $\mathfrak{X}_1$ and $\mathfrak{X}_2$ be two members of $\mathfrak{A}$, and let δ_1 be the lower bound of positive numbers δ such that $\mathfrak{U}(\mathfrak{X}_1, \delta) \supset \mathfrak{X}_2$ and δ_2 be the lower bound of positive numbers δ such that $\mathfrak{U}(\mathfrak{X}_2, \delta) \supset \mathfrak{X}_1$.

We define the distance between $\mathfrak{X}_1$ and $\mathfrak{X}_2$ to be

$$\Delta(\mathfrak{X}_1, \mathfrak{X}_2) = \delta_1 + \delta_2.$$

We have to show that this function is a metric in the usual sense, i.e. that it has the properties

$$\Delta(\mathfrak{X}_1, \mathfrak{X}_2) \geqslant 0; \quad \Delta(\mathfrak{X}_1, \mathfrak{X}_2) = 0 \quad \text{if and only if } \mathfrak{X}_1 \text{ is } \mathfrak{X}_2, \tag{4·1·1}$$

$$\Delta(\mathfrak{X}_1, \mathfrak{X}_2) = \Delta(\mathfrak{X}_2, \mathfrak{X}_1), \tag{4·1·2}$$

$$\Delta(\mathfrak{X}_1, \mathfrak{X}_2) + \Delta(\mathfrak{X}_2, \mathfrak{X}_3) \geqslant \Delta(\mathfrak{X}_1, \mathfrak{X}_3). \tag{4·1·3}$$

To see that (4·1·1) is true we observe that $\Delta(\mathfrak{X}_1, \mathfrak{X}_2)$ is certainly non-negative since it is the sum of two non-negative numbers. Further, if $\Delta(\mathfrak{X}_1, \mathfrak{X}_2) = 0$ then both δ_1 and δ_2 used in defining $\Delta(\mathfrak{X}_1, \mathfrak{X}_2)$ are zero, and this implies that $\mathfrak{X}_1$ is $\mathfrak{X}_2$. The converse of this is clearly true.

The equality (4·1·2) follows from the symmetry of the definition as between $\mathfrak{X}_1$ and $\mathfrak{X}_2$.

To prove (4·1·3) let $\eta_1, \eta_2, \eta_3, \eta_4$ be four positive numbers for which

$$\mathfrak{U}(\mathfrak{X}_1, \eta_1) \supset \mathfrak{X}_2, \quad \mathfrak{U}(\mathfrak{X}_2, \eta_2) \supset \mathfrak{X}_3,$$

$$\mathfrak{U}(\mathfrak{X}_2, \eta_3) \supset \mathfrak{X}_1, \quad \mathfrak{U}(\mathfrak{X}_3, \eta_4) \supset \mathfrak{X}_2.$$

Then
$$\mathfrak{U}(\mathfrak{X}_1, \eta_1 + \eta_2) \supset \mathfrak{X}_3, \quad \mathfrak{U}(\mathfrak{X}_3, \eta_3 + \eta_4) \supset \mathfrak{X}_1,$$

and thus
$$\Delta(\mathfrak{X}_1, \mathfrak{X}_3) \leqslant \eta_1 + \eta_2 + \eta_3 + \eta_4.$$

On taking the lower bounds of $\eta_1, \eta_2, \eta_3, \eta_4$ we obtain

$$\Delta(\mathfrak{X}_1, \mathfrak{X}_3) \leqslant \Delta(\mathfrak{X}_1, \mathfrak{X}_2) + \Delta(\mathfrak{X}_2, \mathfrak{X}_3),$$

and (4·1·3) is true.

Definition. *If a sequence of members of* $\mathfrak{A}$ *is given, say* $\{\mathfrak{X}_i\}$, *then by the phrase '*$\mathfrak{X}_i$ *tends to* $\mathfrak{X}$*' or '*$\mathfrak{X}_i \to \mathfrak{X}$*' is meant that there exists a member* $\mathfrak{X}$ *of* $\mathfrak{A}$ *such that*

$$\Delta(\mathfrak{X}_i, \mathfrak{X}) \to 0 \quad \textit{as} \quad i \to \infty.$$

We are really interested in the subclass $\mathfrak{B}$ of $\mathfrak{A}$ whose members are convex sets. There is the following theorem:

THEOREM 29. *If $\mathfrak{X}_i$ tends to $\mathfrak{X}$ and each $\mathfrak{X}_i$ is a convex subset of $\mathfrak{S}(R)$, then so is $\mathfrak{X}$.*

If $\mathfrak{X}$ is not convex then there are two points $\mathbf{x}_1$ and $\mathbf{x}_2$ of $\mathfrak{X}$ and a point $\mathbf{x}_0$ which is exterior to $\mathfrak{X}$ and which belongs to the segment $\mathfrak{H}(\mathbf{x}_1, \mathbf{x}_2)$. Since $\mathfrak{X}$ is closed there is a positive number δ such that

$$\mathfrak{U}(\mathbf{x}_0, \delta) \cap \mathfrak{X} = \phi.$$

Choose i so large that $\Delta(\mathfrak{X}_i, \mathfrak{X}) < \frac{1}{2}\delta$ and two points $\mathbf{x}_1', \mathbf{x}_2'$ of $\mathfrak{X}_i$ such that

$$|\mathbf{x}_1 - \mathbf{x}_1'| < \tfrac{1}{2}\delta, \quad |\mathbf{x}_2 - \mathbf{x}_2'| < \tfrac{1}{2}\delta.$$

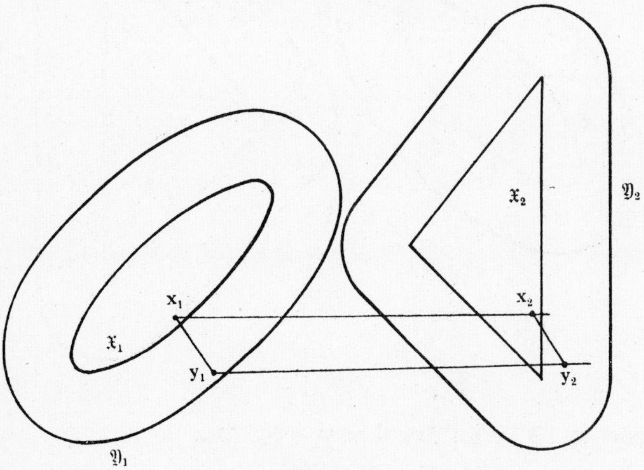

Fig. 8

The segment $\mathfrak{H}(\mathbf{x}_1', \mathbf{x}_2')$ contains a point $\mathbf{x}_0'$ such that

$$|\mathbf{x}_0 - \mathbf{x}_0'| < \tfrac{1}{2}\delta.$$

But then for any $\mathbf{x} \in \mathfrak{X}$ $\quad |\mathbf{x} - \mathbf{x}_0'| > \frac{1}{2}\delta$,

and this implies that $\Delta(\mathfrak{X}_i, \mathfrak{X}) > \frac{1}{2}\delta$. This contradiction shows that $\mathfrak{X}$ is convex.

THEOREM 30. *If $\mathfrak{X}_1, \mathfrak{X}_2$ are given closed convex sets and*

$$\mathfrak{Y}_1 = \overline{[\mathfrak{U}(\mathfrak{X}_1, \delta)]}, \quad \mathfrak{Y}_2 = \overline{[\mathfrak{U}(\mathfrak{X}_2, \delta)]},$$

then
$$\Delta(\mathfrak{Y}_1, \mathfrak{Y}_2) = \Delta(\mathfrak{X}_1, \mathfrak{X}_2).$$

We suppose that R is so large that $\mathfrak{Y}_1$ and $\mathfrak{Y}_2$ belong to $\mathfrak{S}(R)$. Now if $\mathfrak{U}(\mathfrak{X}_1, \epsilon) \supset \mathfrak{X}_2$ then $\mathfrak{U}(\mathfrak{Y}_1, \epsilon) \supset \mathfrak{Y}_2$. For corresponding to any point $\mathbf{y}_2$ of $\mathfrak{Y}_2$ there is a point $\mathbf{x}_2$ of $\mathfrak{X}_2$ such that $|\mathbf{x}_2 - \mathbf{y}_2| \leqslant \delta$. Since $\mathfrak{U}(\mathfrak{X}_1, \epsilon) \supset \mathfrak{X}_2$ there is a point $\mathbf{x}_1$ of $\mathfrak{X}_1$ such that $|\mathbf{x}_1 - \mathbf{x}_2| < \epsilon$. Let $\mathbf{y}_1$ be that point which is so situated that the four points $\mathbf{x}_1, \mathbf{y}_1, \mathbf{y}_2, \mathbf{x}_2$ are the four vertices, in order, of a parallelogram (see fig. 8). Then

$$|\mathbf{x}_1 - \mathbf{y}_1| = |\mathbf{x}_2 - \mathbf{y}_2| \leqslant \delta,$$

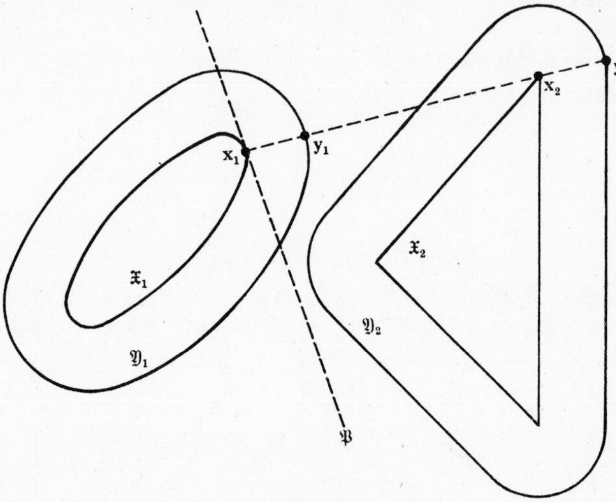

Fig. 9

and since $\mathbf{x}_1 \in \mathfrak{X}_1$ it follows that $\mathbf{y}_1 \in \mathfrak{Y}_1$. Also

$$|\mathbf{y}_1 - \mathbf{y}_2| = |\mathbf{x}_1 - \mathbf{x}_2| < \epsilon,$$

and thus $\mathbf{y}_2 \in \mathfrak{U}(\mathfrak{Y}_1, \epsilon)$. Hence $\mathfrak{U}(\mathfrak{Y}_1, \epsilon) \supset \mathfrak{Y}_2$.

Conversely, if $\mathfrak{U}(\mathfrak{Y}_1, \epsilon) \supset \mathfrak{Y}_2$, then $\overline{[\mathfrak{U}(\mathfrak{X}_1, \epsilon)]} \supset \mathfrak{X}_2$. For if this second inclusion was not true there would exist a point $\mathbf{x}_2$ of $\mathfrak{X}_2$ such that if $\mathbf{x}_1$ is the point of $\mathfrak{X}_1$ nearest to $\mathbf{x}_2$ then $|\mathbf{x}_1 - \mathbf{x}_2| > \epsilon$ (see fig. 9). The hyperplane through $\mathbf{x}_1$ perpendicular to $\mathbf{x}_1\mathbf{x}_2$ is a support hyperplane of $\mathfrak{X}_1$ which we denote by $\mathfrak{P}$. Let $\mathbf{y}_2$ be a point of $\mathfrak{Y}_2$ such that

$$|\mathbf{x}_2 - \mathbf{y}_2| = \delta, \quad \mathbf{x}_2 \in \mathfrak{H}(\mathbf{x}_1, \mathbf{y}_2). \tag{4·1·4}$$

Now $\mathbf{x}_1$ is the point of $\mathfrak{X}_1$ nearest to $\mathbf{y}_2$, and if $\mathbf{y}_1$ is such that

$$|\mathbf{x}_1 - \mathbf{y}_1| = \delta, \quad \mathbf{y}_1 \in \mathfrak{H}(\mathbf{x}_1, \mathbf{y}_2), \tag{4·1·5}$$

then $\mathbf{y}_1$ is the point of $\mathfrak{Y}_1$ nearest to $\mathbf{y}_2$. Since $\mathfrak{U}(\mathfrak{Y}_1, \epsilon) \supset \mathfrak{Y}_2$, $|\mathbf{y}_1 - \mathbf{y}_2| < \epsilon$. Hence from (4·1·5) we have

$$|\mathbf{x}_1 - \mathbf{y}_2| < \epsilon + \delta. \tag{4·1·6}$$

But $|\mathbf{x}_1 - \mathbf{x}_2| > \epsilon$ by assumption, and from (4·1·4) we conclude that

$$|\mathbf{x}_1 - \mathbf{y}_2| > \epsilon + \delta. \tag{4·1·7}$$

Since (4·1·6) and (4·1·7) are in contradiction the result follows.

The theorem now follows from the definition of the metric Δ.

THEOREM 31. *If $\mathfrak{X}_i \to \mathfrak{X}$ as $i \to \infty$ and $\mathfrak{Y}$ is an n-dimensional convex set whose interior meets $\mathfrak{X}$, where $\mathfrak{X}_i, \mathfrak{X}, \mathfrak{Y}$ all belong to $\mathfrak{B}$, then $\mathfrak{X}_i \cap \mathfrak{Y} \to \mathfrak{X} \cap \mathfrak{Y}$.*

We have to show that given $\epsilon > 0$ there exist integers M and N such that

$$\mathfrak{U}(\mathfrak{X} \cap \mathfrak{Y}, \epsilon) \supset \mathfrak{X}_i \cap \mathfrak{Y} \quad (i \geqslant M), \tag{4·1·8}$$

$$\mathfrak{U}(\mathfrak{X}_i \cap \mathfrak{Y}, \epsilon) \supset \mathfrak{X} \cap \mathfrak{Y} \quad (i \geqslant N). \tag{4·1·9}$$

If (4·1·8) is false there is a positive number ϵ, a sequence of integers tending to infinity, $\{i_j\}$, and a sequence of points $\{\mathbf{p}_j\}$ such that $\mathbf{p}_j \in \mathfrak{X}_{i_j} \cap \mathfrak{Y}$ and the distance of $\mathbf{p}_j$ from $\mathfrak{X} \cap \mathfrak{Y}$ is at least ϵ. Further, by selecting a subsequence if necessary, we may suppose that the sequence $\{\mathbf{p}_j\}$ converges to a point $\mathbf{p}$. Since $\mathfrak{Y}$ is closed $\mathbf{p} \in \mathfrak{Y}$. Also, $\mathfrak{X}_i \to \mathfrak{X}$ and $\mathbf{p}_j \in \mathfrak{X}_{i_j}$, thus $\mathbf{p} \in \mathfrak{X}$. Hence, finally, $\mathbf{p} \in \mathfrak{X} \cap \mathfrak{Y}$. But for a sufficiently large integer j, $|\mathbf{p}_j - \mathbf{p}| < \epsilon$ and we have a contradiction since each $\mathbf{p}_j$ is distant at least ϵ from $\mathfrak{X} \cap \mathfrak{Y}$.

Next if (4·1·9) is false there is a sequence of integers $\{i_j\}$ and a sequence of points $\{\mathbf{p}_j\}$ of $\mathfrak{X} \cap \mathfrak{Y}$ such that the distance of $\mathbf{p}_j$ from $\mathfrak{X}_{i_j} \cap \mathfrak{Y}$ is at least ϵ for each j, $j = 1, 2, \ldots$. Again we may suppose that the sequence $\{\mathbf{p}_j\}$ converges to a point $\mathbf{p}$ of $\mathfrak{X} \cap \mathfrak{Y}$. Let $\mathbf{q}$ be a point of the interior of $\mathfrak{Y}$ that also belongs to $\mathfrak{X}$ (see fig. 10). Let $\mathbf{r}$ be a point such that

$$\mathbf{r} \in \mathfrak{H}(\mathbf{p}, \mathbf{q}), \quad |\mathbf{p} - \mathbf{r}| < \tfrac{1}{2}\epsilon.$$

The point $\mathbf{r}$ belongs to both $\mathfrak{X}$ and to the interior of $\mathfrak{Y}$. There is a positive number δ, $\delta < \frac{1}{4}\epsilon$, such that $\mathfrak{S}(\mathbf{r}, \delta) \subset \mathfrak{Y}$. Next there is an integer N_1 such that for $i \geqslant N_1$, $\Delta(\mathfrak{X}_i, \mathfrak{X}) < \delta$, and thus if $i_j \geqslant N_1$ there is a point $\mathbf{r}_j$ of $\mathfrak{X}_{i_j}$ with $|\mathbf{r}_j - \mathbf{r}| < \delta$. Now $\mathbf{r}_j$ belongs both to $\mathfrak{X}_{i_j}$ and to $\mathfrak{Y}$; thus it belongs to $\mathfrak{X}_{i_j} \cap \mathfrak{Y}$, and further its distance from $\mathbf{p}$ is at most $\delta+\frac{1}{2}\epsilon<\frac{3}{4}\epsilon$. We may choose M so large that for $j \geqslant M$, $|\mathbf{p}_j-\mathbf{p}|<\frac{1}{4}\epsilon$. Then for j so large that $j \geqslant M$, $i_j \geqslant N_1$, the distance of $\mathbf{p}_j$ from $\mathbf{r}_j$ is less than ϵ. This contradicts the definition of $\mathbf{p}_j$. Thus the theorem is proved.

Exercises 4·1

1. Show that if $\mathfrak{X}_1, \mathfrak{X}_2, \mathfrak{X}_3$ are three bounded closed sets such that $\mathfrak{X}_1 \subset \mathfrak{X}_2 \subset \mathfrak{X}_3$, then

$$\Delta(\mathfrak{X}_1, \mathfrak{X}_2) \leqslant \Delta(\mathfrak{X}_1, \mathfrak{X}_3), \quad \Delta(\mathfrak{X}_2, \mathfrak{X}_3) \leqslant \Delta(\mathfrak{X}_1, \mathfrak{X}_3).$$

Give an example to show that it is possible to have $\mathfrak{X}_1 \subset \mathfrak{X}_2 \subset \mathfrak{X}_3$, $\mathfrak{X}_1 \neq \mathfrak{X}_2 \neq \mathfrak{X}_3$ and $\Delta(\mathfrak{X}_1, \mathfrak{X}_2)=\Delta(\mathfrak{X}_2, \mathfrak{X}_3)=\Delta(\mathfrak{X}_1, \mathfrak{X}_3)$.

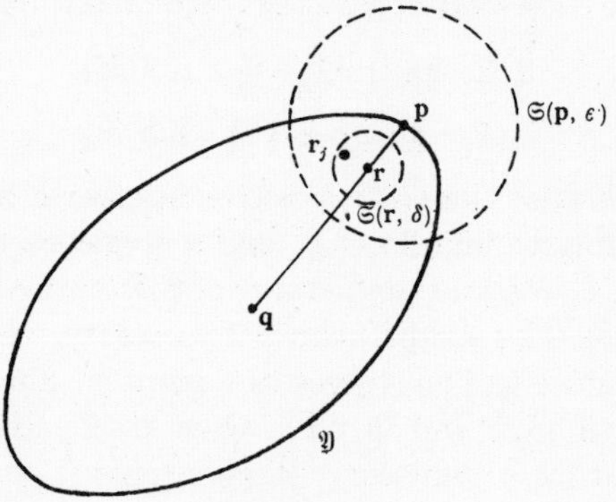

Fig. 10

2. Show that

$$\Delta(\mathfrak{H}(\mathfrak{X}_1, \mathfrak{X}_2), \mathfrak{H}(\mathfrak{Y}_1, \mathfrak{Y}_2)) \leqslant \Delta(\mathfrak{X}_1, \mathfrak{Y}_1)+\Delta(\mathfrak{X}_2, \mathfrak{Y}_2).$$

3. The sequence of convex sets $\{\mathfrak{X}_i\}$ converges to the n-dimensional convex set $\mathfrak{X}$. $\mathbf{p}$ is an interior point of $\mathfrak{X}$ and $\mathfrak{Y}_i$ is the largest set that is directly similar to $\mathfrak{X}_i$ with $\mathbf{p}$ as centre of similitude, and that is contained in $\mathfrak{X}$. Show that $\mathfrak{Y}_i \rightarrow \mathfrak{X}$ and that the ratio of similitude of $\mathfrak{Y}_i$ to $\mathfrak{X}_i$ tends to unity.

2. The Blaschke selection theorem

Theorem 32. *Every infinite aggregate of closed convex subsets of a bounded portion of* R^n, *contains an infinite subsequence that converges to a closed non-void convex subset.*

From the infinite aggregate select an infinite sequence of sets, say, $\mathfrak{X}(1), \mathfrak{X}(2), \ldots, \mathfrak{X}(N), \ldots$. Let δ be a positive number and write

$$\mathfrak{Y}(N) = \overline{[\mathfrak{U}(\mathfrak{X}(N), \delta)]}. \tag{4·2·1}$$

The theorem is proved in two stages as follows:

(i) We show that there is a subsequence of $\{\mathfrak{Y}(N)\}$, say $\{\mathfrak{Y}(N_i)\}$, such that

$$\Delta(\mathfrak{Y}(N_i), \mathfrak{Y}(N_j)) \to 0 \quad \text{as} \quad i, j \to \infty. \tag{4·2·2}$$

(ii) We show that if a sequence of closed convex sets contained in $\mathfrak{S}(R)$, say $\{\mathfrak{Z}_i\}$, is such that

$$\Delta(\mathfrak{Z}_i, \mathfrak{Z}_j) \to 0 \quad \text{as} \quad i, j \to \infty. \tag{4·2·3}$$

then there exists a closed convex set $\mathfrak{Z}$ such that

$$\Delta(\mathfrak{Z}, \mathfrak{Z}_i) \to 0 \quad \text{as} \quad i \to \infty. \tag{4·2·4}$$

Suppose for the moment that (i) and (ii) have been proved, then the theorem can be deduced as follows. (4·2·2) and Theorem 30 imply that there is a subsequence of sets $\{\mathfrak{X}(N_i)\}$ such that

$$\Delta(\mathfrak{X}(N_i), \mathfrak{X}(N_j)) \to 0 \quad \text{as} \quad i, j \to \infty. \tag{4·2·5}$$

But then (ii) applied to (4·2·5) gives the required result.

We next prove the assertions (i) and (ii).

(i) Let $\mathbf{p}_1, \mathbf{p}_2, \ldots$ be a sequence of points dense in $\mathfrak{S}(R)$ and $\{\epsilon_i\}$ be a sequence of positive numbers decreasing to zero, $\epsilon_1 < \frac{1}{2}\delta$. Assume inductively that we have defined

$$N_{i,1}, N_{i,2}, N_{i,3}, \ldots \quad \text{for} \quad i = 0, 1, \ldots, k,$$

as a sequence of integers such that $\{N_{i+1,j}\}$ is a subsequence of $\{N_{i,j}\}$, $N_{0,j} = j$, and if $i > 0$

$$\Delta(\mathfrak{Y}(N_{i,j}), \mathfrak{Y}(N_{i,h})) < \epsilon_i \quad j, h = 1, 2, \ldots. \tag{4·2·6}$$

We shall show that it is possible to select a subsequence of $\{N_{k,j}\}$ say $\{N_{k+1,j}\}$, such that

$$\Delta(\mathfrak{Y}(N_{k+1,j}), \mathfrak{Y}(N_{k+1,h})) < \epsilon_{k+1}. \tag{4·2·7}$$

Choose an integer M, so large that every point of $\mathfrak{S}(R)$ is distant less than $\frac{1}{4}\epsilon_{k+1}$ from some $\mathbf{p}_j$ with $j \leqslant M$. Denote the subset of the points $\mathbf{p}_j, j \leqslant M$, contained in $\mathfrak{Y}(N)$ by $\mathfrak{K}(N)$. Every point of $\mathfrak{U}(\mathfrak{X}(N), \delta - \frac{1}{4}\epsilon_{k+1})$ is distant from one of the points $\mathbf{p}_j, j \leqslant M$, by at most $\frac{1}{4}\epsilon_{k+1}$, and this particular $\mathbf{p}_j$ must be contained in $\mathfrak{Y}(N)$; thus it must belong to $\mathfrak{K}(N)$. Hence

$$\mathfrak{U}(\mathfrak{K}(N), \tfrac{1}{4}\epsilon_{k+1}) \supset \mathfrak{U}(\mathfrak{X}(N), \delta - \tfrac{1}{4}\epsilon_{k+1}). \qquad (4\cdot2\cdot8)$$

Thus
$$\mathfrak{U}(\mathfrak{K}(N), \tfrac{1}{2}\epsilon_{k+1}) \supset \mathfrak{U}(\mathfrak{X}(N), \delta) = \mathfrak{Y}(N)^0. \qquad (4\cdot2\cdot9)$$

But
$$\mathfrak{Y}(N) \supset \mathfrak{K}(N),$$

and thus with (4·2·9) we have

$$\Delta(\mathfrak{K}(N), \mathfrak{Y}(N)) < \tfrac{1}{2}\epsilon_{k+1}. \qquad (4\cdot2\cdot10)$$

The finite set of points $\mathbf{p}_1, \mathbf{p}_2, \ldots, \mathbf{p}_M$ can be arranged into subsets in at most a finite number of ways, so that from the infinite sequence $\mathfrak{Y}(N_{k,j})$ we can select an infinite subsequence $\mathfrak{Y}(N_{k+1,j})$ such that for each member of this subsequence the set $\mathfrak{K}(N_{k+1,j})$ is the same set. Then, by the inequality (4·1·3), we have, from (4·2·10),

$$\Delta(\mathfrak{Y}(N_{k+1,j}), \mathfrak{Y}(N_{k+1,h})) < \epsilon_{k+1}.$$

Thus the next step in the inductive definition of the sequence of sequences of integers $\{N_{i,j}\}$ is completed. Consider the sequence

$$N_{1,1}, N_{2,2}, \ldots, N_{h,h}, \ldots.$$

For $h \geqslant i$, $N_{h,h}$ is an element of the sequence $\{N_{i,j}\}$. Thus

$$\Delta(\mathfrak{Y}(N_{h,h}), \mathfrak{Y}(N_{j,j})) \to 0 \quad \text{as} \quad h, j \to \infty,$$

and this completes the proof of (i).

(ii) To prove the second assertion we write

$$\mathfrak{Z} = \bigcap_{i=1}^{\infty} \left[\overline{\bigcup_{j=i}^{\infty} \mathfrak{Z}_j}\right].$$

We shall show that $\mathfrak{Z}$ is the limit of the sequence $\mathfrak{Z}_i$.

Given $\epsilon > 0$ there is an integer N such that

$$\Delta(\mathfrak{Z}_i, \mathfrak{Z}_j) < \epsilon \quad (i, j \geqslant N). \qquad (4\cdot2\cdot11)$$

Then
$$\bigcup_{j=i}^{\infty} \mathfrak{Z}_j \subset \mathfrak{U}(\mathfrak{Z}_i, \epsilon).$$

Thus
$$\mathfrak{Z} \subset \left[\overline{\bigcup_{j=i}^{\infty} \mathfrak{Z}_j}\right] \subset \mathfrak{U}(\mathfrak{Z}_i, 2\epsilon). \qquad (4\cdot2\cdot12)$$

Let $\mathbf{z}_0$ be any point of $\mathfrak{Z}_i$. Write

$$\overline{[\mathfrak{U}(\mathbf{z}_0, \epsilon)]} \cap \mathfrak{Z}_j = W_j \quad (i, j \geq N).$$

W_j is closed and by (4·2·11) is non-void. Thus the set

$$\mathfrak{T}_N = \bigcap_{i=N}^{\infty} \overline{\left[\bigcup_{j=i}^{\infty} W_j\right]}$$

is also closed and non-void. Now $\mathfrak{T}_N$ is contained in both $\overline{[\mathfrak{U}(\mathbf{z}_0, \epsilon)]}$ and in $\mathfrak{Z}$. Thus to any point $\mathbf{z}_0$ of $\mathfrak{Z}_i$, $i \geq N$, there corresponds a point of $\mathfrak{Z}$ whose distance from $\mathbf{z}_0$ is at most ϵ. That is to say

$$\mathfrak{Z}_i \subset \mathfrak{U}(\mathfrak{Z}, 2\epsilon) \quad (i \geq N). \tag{4·2·13}$$

Thus, finally, combining (4·2·12) and (4·2·13), we have

$$\Delta(\mathfrak{Z}, \mathfrak{Z}_i) < 4\epsilon \quad (i \geq N),$$

and this completes the proof of (ii).

Thus the theorem is established.

EXERCISES 4·2

1. A sequence of support hyperplanes $\{\mathfrak{Q}_i\}$ to a convex set $\mathfrak{X}$ is given and $\mathfrak{X} \subset \mathfrak{S}(R)$. Show that if the sequence $\mathfrak{Q}_i \cap \overline{[\mathfrak{S}(R)]}$ converges, then so does $\mathfrak{Q}_i \cap \overline{[\mathfrak{S}(r)]}$ for any $r > R$, and that the limit of this last sequence, say $\mathfrak{T}(r)$, is the part of a support hyperplane of $\mathfrak{X}$ contained in $\overline{[\mathfrak{S}(r)]}$.

2. $\mathfrak{X}$ is a convex set with interior points one of which is $\mathbf{p}$. $\mathbf{x}_f$ is a variable point of the frontier of $\mathfrak{X}$. Show that the set of angles formed by $\mathbf{p}\mathbf{x}_f$ with normals to support hyperplanes at $\mathbf{x}_f$ is closed when $\mathbf{x}_f$ is fixed and that it does not contain $\frac{1}{2}\pi$. Show that the same is true when $\mathbf{x}_f$ is allowed to vary in $\operatorname{Fr}\mathfrak{X}$.

3. Approximations by convex polytopes and regular convex sets

In many problems it is convenient to establish the required result for a special class of convex sets and then to extend it to all other convex sets by an approximation argument. There are two classes of convex sets that are important in this connexion. They are: (i) the convex polytopes, and (ii) the regular convex sets. The definition of this second class of convex sets has been given in § 1·11.

We shall show that we can approximate to a given convex set by means of either a convex polytope or a regular convex set. Another way of stating these results is that the convex polytopes and the regular convex sets are dense in $\mathfrak{B}$, the space of all convex subsets of $\mathfrak{S}(R)$.

THEOREM 33. *If $\mathfrak{X}$ is a bounded closed convex set and if ϵ is a given positive number, there are convex polytopes $\mathfrak{P}_1, \mathfrak{P}_2$ such that*

$$\mathfrak{P}_1 \subset \mathfrak{X} \subset \mathfrak{P}_2 \quad \textit{and} \quad \Delta(\mathfrak{P}_1, \mathfrak{P}_2) < \epsilon.$$

We shall assume that $\mathfrak{X}$ is n-dimensional. If this were not the case we should use $\mathfrak{L}(\mathfrak{X})$ in place of R^n.

Suppose that $\mathfrak{X} \subset \mathfrak{S}(R)$ and choose a finite set of points $\mathbf{p}_1, \mathbf{p}_2, \ldots, \mathbf{p}_M$ in $\mathfrak{S}(R)$ such that if $\mathfrak{T} = \bigcup_{i=1}^{M} \mathbf{p}_i$, then

$$\mathfrak{U}(\mathfrak{T}, \tfrac{1}{4}\eta) \supset \mathfrak{S}(R),$$

where η is a positive number which will be defined later in terms of ϵ. Write

$$\mathfrak{P}_2 = \mathfrak{H}(\mathfrak{T} \cap \mathfrak{U}(\mathfrak{X}, \eta)),$$

then clearly since $\mathfrak{U}(\mathfrak{X}, \eta)$ is convex (though not closed),

$$\mathfrak{U}(\mathfrak{X}, \eta) \supset \mathfrak{P}_2. \tag{4·3·1}$$

Further

$$\mathfrak{P}_2 \supset \mathfrak{X}, \tag{4·3·2}$$

for if this were not true there would be a point $\mathbf{x}$ of $\mathfrak{X} \dot{-} \mathfrak{P}_2$. There is a hyperplane $\mathfrak{Q}$ which passes through $\mathbf{x}$ and does not cut $\mathfrak{P}_2$. Denote the two points which lie in the line through $\mathbf{x}$ perpendicular to $\mathfrak{Q}$ at a distance $\frac{1}{2}\eta$ from $\mathbf{x}$ by $\mathbf{x}_1$ and $\mathbf{x}_2$. There are points $\mathbf{p}_{i_1}, \mathbf{p}_{i_2}$ of $\mathfrak{T}$ such that

$$|\mathbf{p}_{i_1} - \mathbf{x}_1| < \tfrac{1}{4}\eta, \quad |\mathbf{p}_{i_2} - \mathbf{x}_2| < \tfrac{1}{4}\eta.$$

But $\mathbf{p}_{i_1}$ and $\mathbf{p}_{i_2}$ lie on opposite sides of $\mathfrak{Q}$ and both belong to $\mathfrak{P}_2$. This is a contradiction as it implies that $\mathfrak{Q}$ cuts $\mathfrak{P}_2$. Hence (4·3·2) is true, and combining with (4·3·1) we obtain

$$\Delta(\mathfrak{P}_2, \mathfrak{X}) < \eta. \tag{4·3·3}$$

Next choose a point $\mathbf{x}_i$ of the interior of $\mathfrak{X}$ and consider a half-ray which terminates at $\mathbf{x}_i$ and which meets the frontier of $\mathfrak{X}$

at $\mathbf{x}_f$. There is a positive number θ such that the angle between the outward normal perpendicular to a support hyperplane at $\mathbf{x}_f$ and the line $\mathbf{x}_i\mathbf{x}_f$ produced is less than $\frac{1}{2}\pi-\theta$. Further, we can choose θ to be the same whatever half-line is considered. Suppose that $\mathbf{x}_i\mathbf{x}_f$ produced meets the frontier of $\mathfrak{P}_2$ in $\mathbf{p}$. The perpendicular distance from $\mathbf{p}$ to a support hyperplane to $\mathfrak{X}$ at $\mathbf{x}_f$ is at most η. Thus

$$|\mathbf{x}_f-\mathbf{p}|<\eta/\sin\theta.$$

Now $\mathbf{x}_i$ is an interior point of $\mathfrak{X}$, and thus there is a positive number δ such that for all $\mathbf{x}_f \in \operatorname{Fr}\mathfrak{X}$

$$|\mathbf{x}_i-\mathbf{x}_f|>\delta.$$

Hence
$$|\mathbf{x}_i-\mathbf{p}|/|\mathbf{x}_i-\mathbf{x}_f|<1+\eta/(\delta\sin\theta). \tag{4·3·4}$$

Let D be the diameter of $\mathfrak{X}$ and choose η small and positive such that

$$\eta<\tfrac{1}{2}\epsilon,\quad D\eta/(\delta\sin\theta)<\tfrac{1}{2}\epsilon. \tag{4·3·5}$$

Let $\mathfrak{P}_1$ be the polytope similar to $\mathfrak{P}_2$ with $\mathbf{x}_i$ as centre of similitude and reduced in the ratio $1:1+\eta/(\delta\sin\theta)$. By (4·3·4) we have

$$\mathfrak{P}_1\subset\mathfrak{X}. \tag{4·3·6}$$

Further, if the segment $\mathbf{x}_i\mathbf{x}_f$ meets the frontier of $\mathfrak{P}_1$ in $\mathbf{p}_1$ then

$$|\mathbf{x}_f-\mathbf{p}_1|<D[1-(1+\eta/(\delta\sin\theta))^{-1}]<\tfrac{1}{2}\epsilon.$$

Hence
$$\Delta(\mathfrak{X},\mathfrak{P}_1)<\tfrac{1}{2}\epsilon. \tag{4·3·7}$$

From (4·3·3), (4·3·5) and (4·3·7) it follows that

$$\Delta(\mathfrak{P}_1,\mathfrak{P}_2)<\epsilon.$$

Thus the proof of the theorem is complete.

We can prove an analogous result with regular convex sets in place of convex polytopes provided that $\mathfrak{X}$ has interior points. In any case whether $\mathfrak{X}$ has interior points or not we have the following theorem:

THEOREM 34. *If $\mathfrak{X}$ is a convex set and η is a given positive number there is regular convex set $\mathfrak{R}$ such that*

$$\mathfrak{X}\subset\mathfrak{R},\quad \Delta(\mathfrak{X},\mathfrak{R})<\eta.$$

Suppose that $\mathfrak{X} \subset \mathfrak{S}(R)$ and denote by $\mathfrak{X}(r)$ the intersection of all closed spheres of radius r, $r > R$, that contain $\mathfrak{X}$. $\mathfrak{X}(r)$ is a convex set containing $\mathfrak{X}$. We show that as $r \to \infty$, $\Delta(\mathfrak{X}(r), \mathfrak{X}) \to 0$. Now $\mathfrak{X}(r) \supset \mathfrak{X}$, and if $r_1 < r_2$, $\mathfrak{X}(r_1) \supset \mathfrak{X}(r_2)$; thus if the relation $\Delta(\mathfrak{X}(r), \mathfrak{X}) \to 0$ were not correct, there would be a point $\mathbf{x}_e$ belonging to every $\mathfrak{X}(r)$ and at a positive distance from $\mathfrak{X}$ (see fig. 11). Suppose this distance is ϵ. Let D be the diameter of $\mathfrak{X}$ and let $\mathbf{x}$ be the point of $\mathfrak{X}$ nearest to $\mathbf{x}_e$. On $\mathbf{x}_e\mathbf{x}$ produced let $\mathbf{c}$ be a point such that

$$|\mathbf{x}-\mathbf{c}| = r_1 - \epsilon_1, \quad \mathbf{x} \in \mathfrak{H}(\mathbf{x}_e, \mathbf{c}) \quad (0 < \epsilon_1 < \epsilon),$$

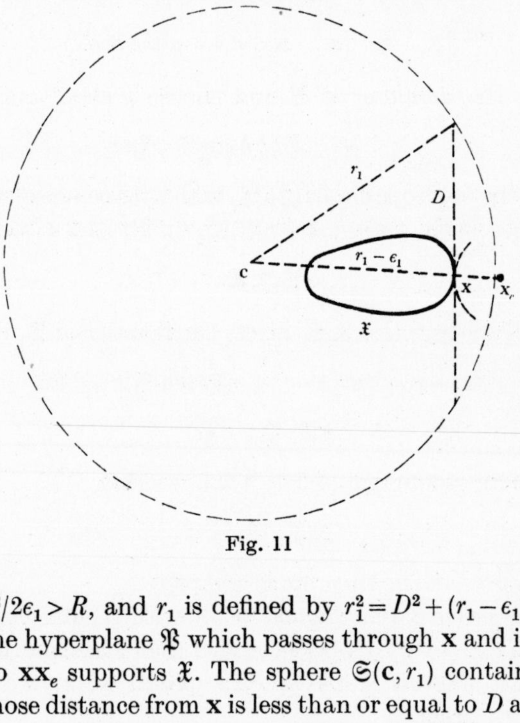

Fig. 11

where $D^2/2\epsilon_1 > R$, and r_1 is defined by $r_1^2 = D^2 + (r_1 - \epsilon_1)^2$

Now the hyperplane $\mathfrak{P}$ which passes through $\mathbf{x}$ and is perpendicular to $\mathbf{x}\mathbf{x}_e$ supports $\mathfrak{X}$. The sphere $\mathfrak{S}(\mathbf{c}, r_1)$ contains all the points whose distance from $\mathbf{x}$ is less than or equal to D and which lie on the side of $\mathfrak{P}$ opposite to $\mathbf{x}_e$, and $r_1 > R$. Thus

$$\mathfrak{S}(\mathbf{c}, r_1) \supset \mathfrak{X}, \quad \text{and} \quad \mathfrak{S}(\mathbf{c}, r_1) \supset \mathfrak{X}(r_1)$$

and since $\mathfrak{S}(\mathbf{c}, r_1) \not\ni \mathbf{x}_e$ we have a contradiction and $\Delta(\mathfrak{X}(r), \mathfrak{X}) \to 0$ as $r \to \infty$.

Hence we can choose r so large that

$$\Delta(\mathfrak{X}(r), \mathfrak{X}) < \tfrac{1}{2}\eta.$$

Write $\mathfrak{R} = \overline{\mathfrak{U}(\mathfrak{X}(r), \tfrac{1}{2}\eta)}$, then

$$\mathfrak{X} \subset \mathfrak{R}, \quad \Delta(\mathfrak{R}, \mathfrak{X}) < \eta,$$

and we have to show that $\mathfrak{R}$ is a regular convex set in order to complete the proof of the theorem.

First, if $\mathbf{q}_f$ is a frontier point of $\mathfrak{R}$, there is a point $\mathbf{x}$ of $\mathfrak{X}(r)$ such that $|\mathbf{q}_f - \mathbf{x}| \leqslant \tfrac{1}{2}\eta$. Now

$$\mathfrak{S}(\mathbf{x}, \tfrac{1}{2}\eta) \subset \mathfrak{R},$$

and there is a unique tangent hyperplane to $\mathfrak{S}(\mathbf{x}, \tfrac{1}{2}\eta)$ at $\mathbf{q}_f$. Hence the support hyperplane to $\mathfrak{R}$ at $\mathbf{q}_f$ is unique.

Next if a support hyperplane $\mathfrak{Q}$ of $\mathfrak{R}$ met the frontier of $\mathfrak{R}$ in two distinct points $\mathbf{q}_1$ and $\mathbf{q}_2$, then there would exist points $\mathbf{x}_1$ and $\mathbf{x}_2$ of $\mathfrak{X}(r)$ such that

$$|\mathbf{x}_1 - \mathbf{q}_1| = |\mathbf{x}_2 - \mathbf{q}_2| = \tfrac{1}{2}\eta.$$

Further, $\mathbf{q}_1\mathbf{x}_1$ and $\mathbf{q}_2\mathbf{x}_2$ are both perpendicular to the support hyperplane $\mathfrak{Q}$. Now $\mathfrak{X}(r)$ is an intersection of spheres of radius r and therefore contains the intersection of all spheres of radius r that contain both $\mathbf{x}_1$ and $\mathbf{x}_2$. But this means that the points of $\mathfrak{H}(\mathbf{x}_1, \mathbf{x}_2)$ other than $\mathbf{x}_1$ and $\mathbf{x}_2$ are interior points of $\mathfrak{X}(r)$. Thus the points of $\mathfrak{H}(\mathbf{q}_1, \mathbf{q}_2)$ other than the end-points are interior points of $\mathfrak{R}$. This is impossible as $\mathfrak{H}(\mathbf{q}_1, \mathbf{q}_2)$ lies on a support hyperplane to $\mathfrak{R}$. Thus every support hyperplane to $\mathfrak{R}$ meets $\mathfrak{R}$ in a single point. The theorem is proved.

Exercise 4·3

1. Show that there is an enumerable sequence of convex sets $\mathfrak{X}_1, \mathfrak{X}_2, \mathfrak{X}_3, \ldots$ such that if $\mathfrak{X}$ is any closed bounded convex set there exists a subsequence $\mathfrak{X}_{i_j}$ such that $\mathfrak{X}_{i_j} \to \mathfrak{X}$ as $j \to \infty$.

4. The volume of a convex set. Continuity

Having defined a metric topology over the class of convex subsets of $\mathfrak{S}(R)$ we next consider certain functions of convex sets that are continuous in this topology. By far the most important of these is the volume of a convex set. But first we define what is meant by continuity in $\mathfrak{B}$.

DEFINITION. *Suppose that corresponding to every member* $\mathfrak{X}$ *of* $\mathfrak{B}$ *a real number* $f(\mathfrak{X})$ *is defined. Then* $f(\mathfrak{X})$ *is said to be continuous at* $\mathfrak{X}_0$ *if for every sequence* $\{\mathfrak{X}_i\}$ *for which* $\mathfrak{X}_i \to \mathfrak{X}_0$ *we have also* $f(\mathfrak{X}_i) \to f(\mathfrak{X}_0)$.

Remarks. (i) It is only necessary for $f(\mathfrak{X})$ to be defined in a neighbourhood of $\mathfrak{X}_0$, i.e. all $\mathfrak{X}$ for which, for some $\epsilon > 0$,

$$\Delta(\mathfrak{X}, \mathfrak{X}_0) < \epsilon.$$

(ii) If $f(\mathfrak{X})$ is continuous for all $\mathfrak{X}_0$ of $\mathfrak{B}$ we say simply that $f(\mathfrak{X})$ is continuous.

It should be noted that the role of the class $\mathfrak{B}$ is to confine our attention to a metric space which is compact and thus in which the Blaschke selection theorem holds. It is essential, in any given problem, to work inside one fixed class $\mathfrak{B}$, but it is usually trivial to find a class $\mathfrak{B}$ inside which it is possible to work. This is because the properties of convex sets with which we are concerned are usually congruence invariant or affine invariant, and if, in the course of our argument, we are led to consider a convex set that does not belong to $\mathfrak{B}$, then we can replace it by one which is congruent or affinely equivalent and which does belong to $\mathfrak{B}$.

DEFINITION. *The volume of a convex set* $\mathfrak{X}$ *is its n-dimensional Lebesgue measure and is denoted by* $V(\mathfrak{X})$.

If $\mathfrak{X}$ *is r-dimensional we shall use* $V_r(\mathfrak{X})$ *for its r-dimensional Lebesgue measure.*

THEOREM 35. *The function* $V(\mathfrak{X})$ *is continuous.*

We have to show that if

$$\mathfrak{X}_i \to \mathfrak{X} \quad \text{as} \quad i \to \infty,$$

then
$$V(\mathfrak{X}_i) \to V(\mathfrak{X}).$$

Now given a positive number ϵ there is an integer M such that

$$\mathfrak{U}(\mathfrak{X}, \epsilon) \supset \mathfrak{X}_i, \quad \mathfrak{U}(\mathfrak{X}_i, \epsilon) \supset \mathfrak{X} \quad \text{for} \quad i \geqslant M.$$

By the argument used in Theorem 33 if $\mathfrak{X}$ has a non-void interior there is a positive number δ and a set similar to $\mathfrak{X}$ but expanded in the ratio $1+\delta : 1$, say $\mathfrak{Y}$, such that $\mathfrak{Y} \supset \mathfrak{U}(\mathfrak{X}, \epsilon)$. Also we can take $\delta = o(1)$ as $\epsilon \to 0$. Thus for $i \geqslant M$

$$V(\mathfrak{X}_i) \leqslant V(\mathfrak{U}(\mathfrak{X}, \epsilon)) \leqslant V(\mathfrak{Y}) = (1+\delta)^n V(\mathfrak{X}). \qquad (4 \cdot 4 \cdot 1)$$

If $V(\mathfrak{X})=0$, $\mathfrak{X}$ lies in an $(n-1)$-dimensional space $\mathfrak{L}$ and if $\Delta(\mathfrak{X}_i, \mathfrak{X})<\eta$ then $V(\mathfrak{X}_i)<2\eta K$ where K is the $(n-1)$-dimensional volume of an $(n-1)$-dimensional sphere of radius R. (All the $\mathfrak{X}_i$ lie in $\mathfrak{S}(R)$). Hence $\lim V(\mathfrak{X}_i)=0$.

If $V(\mathfrak{X})>0$ we need a lower bound for $V(\mathfrak{X}_i)$ which may be obtained as follows. First we show that the frontier of $\mathfrak{X}$ has zero volume. Let $\mathbf{p}$ be an interior point of $\mathfrak{X}$ and $\mathfrak{X}(\lambda)$, $\mathfrak{X}(-\lambda)$ be the sets homothetic to $\mathfrak{X}$ with $\mathbf{p}$ as centre of similitude and $1+\lambda:1$, $1-\lambda:1$ as ratios of similitude, $0<\lambda<1$. Then the frontier of $\mathfrak{X}$ is contained in $\mathfrak{X}(\lambda)\dot{-}\mathfrak{X}(-\lambda)$, and thus has volume not greater than

$$[(1+\lambda)^n-(1-\lambda)^n]\,V(\mathfrak{X}).$$

Since we may choose λ as near 0 as we please, provided $\lambda>0$, it follows that the frontier of $\mathfrak{X}$ has zero n-dimensional measure. Next let $\{\mathbf{p}_i\}$ be a sequence of points dense in the interior of $\mathfrak{X}$. Let $\mathfrak{K}_N$ be the convex cover of the first N of these points. The sequence $\{\mathfrak{K}_N\}$ is a sequence of sets which increases to the interior of $\mathfrak{X}$. Thus

$$\lim_{N\to\infty} V(\mathfrak{K}_N)=V(\mathfrak{X}^0)=V(\mathfrak{X}), \tag{4·4·2}$$

so given $\epsilon>0$ there is an integer N such that

$$V(\mathfrak{K}_N)>V(\mathfrak{X})-\epsilon,$$

Also there is an integer M such that

$$\mathfrak{X}_i\supset\mathfrak{K}_N \quad (i\geqslant M). \tag{4·4·3}$$

The equation (4·4·2) is a well-known measure property and (4·4·3) follows from the fact that if δ is the least of the distances of the points of $\mathfrak{K}_N$ from the frontier of $\mathfrak{X}$, and if $\mathfrak{Y}$ is a convex set not containing $\mathfrak{K}_N$, then

$$\Delta(\mathfrak{Y},\mathfrak{X})\geqslant\delta.$$

From (4·4·2), (4·4·3) and (4·4·1) it follows that

$$V(\mathfrak{X}_i)\to V(\mathfrak{X}) \quad \text{as} \quad i\to\infty.$$

Another property of volume which we shall require later is the following:

THEOREM 36. *If $\mathfrak{X}$ is an $(n-2)$-dimensional convex set then $V(\mathfrak{U}(\mathfrak{X},\delta))=O(\delta^2)$ as δ tends to zero through positive values.*

We can choose a coordinate system so that if $\mathbf{x} \in \mathfrak{X}$, then

$$x_1 = x_2 = 0, \quad |x_\nu| \leqslant K \quad (\nu = 3, \ldots, n).$$

Then if $\mathbf{y} \in \mathfrak{U}(\mathfrak{X}, \delta)$ we have

$$|y_1| \leqslant \delta, \quad |y_2| \leqslant \delta, \quad |y_\nu| \leqslant K + \delta \quad (\nu = 3, \ldots, n).$$

Thus $$V(\mathfrak{U}(\mathfrak{X}, \delta)) \leqslant 4\delta^2(2K + 2\delta)^{n-2} = O(\delta^2),$$

and the theorem is proved.

We prove next a formula for the volume of a convex polytope.

THEOREM 37. *Let the faces of a convex polytope $\mathfrak{X}$ lie in the distinct hyperplanes $\mathfrak{Q}_1, \mathfrak{Q}_2, \ldots, \mathfrak{Q}_k$ and let the perpendicular distance from the origin to $\mathfrak{Q}_i$ be h_i, where h_i is positive if the origin and $\mathfrak{X}$ lie on the same side of $\mathfrak{Q}_i$ and negative in the contrary case. Then*

$$V(\mathfrak{X}) = \frac{1}{n} \sum_{i=1}^{k} h_i . V_{n-1}(\mathfrak{Q}_i \cap \mathfrak{X}). \tag{4·4·4}$$

Let $\mathbf{O}$ be the origin and let $\mathbf{O}'$ be a point of the relative interior of $\mathfrak{X}$. Let h'_i be the perpendicular distance from $\mathbf{O}'$ to $\mathfrak{Q}_i$ (each h'_i is positive or zero). Write $\mathbf{t}$ for the vector from $\mathbf{O}$ to $\mathbf{O}'$ and let $\boldsymbol{\zeta}_i$ be the unit vector perpendicular to $\mathfrak{Q}_i$ directed from the relative interior of $\mathfrak{X}$ to its exterior. Then

$$h'_i = h_i - \mathbf{t} . \boldsymbol{\zeta}_i.$$

Now $$\sum_{i=1}^{k} \mathbf{t} . \boldsymbol{\zeta}_i V_{n-1}(\mathfrak{Q}_i \cap \mathfrak{X}) = 0, \tag{4·4·5}$$

for the expression on the left-hand side of (4·4·5) is the value of the projection of the surface area of $\mathfrak{X}$ in the direction of $\mathbf{t}$ multiplied by $|\mathbf{t}|$, and this is zero, since the projections of the different faces are signed. Every point of the projection belongs to one positive and to one negative face or to the projection of a set of dimension $n-2$ (points where two or more faces of $\mathfrak{X}$ intersect). Thus the expression (4·4·4) is independent of the choice of origin and we shall suppose that the origin is a relative interior point of $\mathfrak{X}$.

If $\mathfrak{X}$ is not n-dimensional then either $\mathfrak{Q}_i \cap \mathfrak{X}$ is at most $(n-2)$-dimensional and $V_{n-1}(\mathfrak{Q}_i \cap \mathfrak{X}) = 0$, or $\mathfrak{Q}_i = \mathfrak{L}(\mathfrak{X})$ and $h_i = 0$. Thus the expression on the right-hand side of (4·4·4) is zero and the theorem is proved.

Suppose that $\mathfrak{X}$ is n-dimensional and the origin is the interior point $\mathbf{p}$ of $\mathfrak{X}$. Then

$$\mathfrak{X} = \bigcup_{i=1}^{k} \mathfrak{H}(\mathbf{p}, \mathfrak{Q}_i \cap \mathfrak{X}). \tag{4·4·6}$$

For since $\mathfrak{X} \supset \mathbf{p} \cup (\mathfrak{Q}_i \cap \mathfrak{X})$ it follows that

$$\mathfrak{X} \supset \bigcup_{i=1}^{k} \mathfrak{H}(\mathbf{p}, \mathfrak{Q}_i \cap \mathfrak{X}).$$

Further, if $\mathbf{x}$ is any point of $\mathfrak{X}$ other than $\mathbf{p}$, the line $\mathbf{px}$ produced meets the frontier of $\mathfrak{X}$ and therefore at least one of the sets $\mathfrak{Q}_i \cap \mathfrak{X}$. Thus $\mathbf{x}$ belongs to at least one of the sets $\mathfrak{H}(\mathbf{p}, \mathfrak{Q}_i \cap \mathfrak{X})$ and the identity (4·4·6) is established.

Next we have

$$\mathfrak{H}(\mathbf{p}, \mathfrak{Q}_i \cap \mathfrak{X}) \cap \mathfrak{H}(\mathbf{p}, \mathfrak{Q}_j \cap \mathfrak{X}) \subset \mathfrak{H}(\mathbf{p}, \mathfrak{Q}_i \cap \mathfrak{Q}_j \cap \mathfrak{X}). \tag{4·4·7}$$

For let $\mathbf{x}$ be any point other than $\mathbf{p}$ of the left-hand side of (4·4·7). The line $\mathbf{px}$ produced meets the frontier of $\mathfrak{X}$ in a unique point $\mathbf{y}$ such that

$$\mathbf{x} \in \mathfrak{H}(\mathbf{p}, \mathbf{y}).$$

Then $\mathbf{y}$ must belong to $\mathfrak{Q}_i$ and to $\mathfrak{Q}_j$. Thus $\mathbf{y} \in \mathfrak{Q}_i \cap \mathfrak{Q}_j$ and the inclusion (4·4·7) follows.

Since $\mathfrak{Q}_i \cap \mathfrak{Q}_j$ is $(n-2)$-dimensional the sets $\mathfrak{H}(\mathbf{p}, \mathfrak{Q}_i \cap \mathfrak{X})$ have at most $(n-1)$-dimensional intersections with each other. Hence

$$V(\mathfrak{X}) = \sum_{i=1}^{k} V(\mathfrak{H}(\mathbf{p}, \mathfrak{Q}_i \cap \mathfrak{X})). \tag{4·4·8}$$

Denote the hyperplane parallel to $\mathfrak{Q}_i$ at a distance χ from $\mathbf{p}$ and lying between $\mathbf{p}$ and $\mathfrak{Q}_i$ by $\mathfrak{Q}_i(\chi)$. Then

$$V_{n-1}[\mathfrak{Q}_i(\chi) \cap \mathfrak{H}(\mathbf{p}, \mathfrak{Q}_i \cap \mathfrak{X})] = (\chi/h_i)^{n-1} V_{n-1}(\mathfrak{Q}_i \cap \mathfrak{X}).$$

Thus

$$V(\mathfrak{H}(\mathbf{p}, \mathfrak{Q}_i \cap \mathfrak{X})) = \int_0^{h_i} \left(\frac{\chi}{h_i}\right)^{n-1} V_{n-1}(\mathfrak{Q}_i \cap \mathfrak{X})\, d\chi = \frac{1}{n} h_i \,.\, V_{n-1}(\mathfrak{Q}_i \cap \mathfrak{X}).$$

From (4·4·8) we have

$$V(\mathfrak{X}) = \frac{1}{n} \sum_{i=1}^{k} h_i \,.\, V_{n-1}(\mathfrak{Q}_i \cap \mathfrak{X}), \tag{4·4·9}$$

and the theorem is proved.

EXERCISE 4·4

1. Denote by $\mathfrak{X}$ the class of all polytopes of unit volume whose faces are perpendicular to some or all of the N directions $\mathbf{l}_1, \mathbf{l}_2, \ldots, \mathbf{l}_N$. It is supposed that $\mathfrak{X}$ is not empty. Show that if $\nu_1, \ldots, \nu_N$ are N positive numbers for which

$$\sum_{i=1}^{N} \nu_i \mathbf{l}_i = \mathbf{O},$$

there is a member of $\mathfrak{X}$ for which the function $\sum_{i=1}^{N} \nu_i p_i$, where p_i is the perpendicular distance of the face perpendicular to $\mathbf{l}_i$ from the origin, assumes a minimum value.

Deduce that there exists a polytope whose N faces are perpendicular to the N directions $\mathbf{l}_i$ and which is such that the face perpendicular to $\mathbf{l}_i$ has $(n-1)$-dimensional measure equal to ν_i.

5. Sets and numbers associated with a convex set

We have already defined two numbers associated with a convex set, its diameter and its volume. In this paragraph we introduce other numbers and sets in terms of which the convex set can be given a more or less precise description.

(i) *The insphere and the circumsphere of* $\mathfrak{X}$

If $\mathfrak{X}$ is a closed bounded convex set the class of spheres contained in $\mathfrak{X}$ have radii which form a bounded set of numbers. By the Blaschke selection theorem there is a sphere of this class whose radius is equal to the upper bound of this set. We use here the facts that the limit of a convergent sequence of spheres is either a sphere or a point and that the radius of a sphere is continuous defined over the class of spheres with the same metric as in § 1. Such a sphere is called an insphere of $\mathfrak{X}$ and its radius the inradius of $\mathfrak{X}$. Although there may be more than one insphere, the inradius is unique. Similarly, a sphere containing $\mathfrak{X}$ and of minimal radius exists. It is unique and is called the circumsphere. Its radius is the circumradius of $\mathfrak{X}$. We denote the inradius, the circumradius, an insphere and the circumsphere of $\mathfrak{X}$ by $r(\mathfrak{X})$, $R(\mathfrak{X})$, $\mathfrak{J}(\mathfrak{X})$ and $\mathfrak{C}(\mathfrak{X})$ respectively.

If $\{\mathfrak{X}_i\}$ is a sequence such that $\mathfrak{X}_i \to \mathfrak{X}$, then $\mathfrak{C}(\mathfrak{X}_i) \to \mathfrak{C}(\mathfrak{X})$, and amongst the $\mathfrak{J}(\mathfrak{X}_i)$ there is a convergent sequence that converges to one of the spheres $\mathfrak{J}(\mathfrak{X})$.

(ii) *The width of a convex set* $\mathfrak{X}$

There are two support hyperplanes of $\mathfrak{X}$ that are perpendicular to a given direction $\mathbf{l}$. Let the distance apart of these two hyperplanes be written $w(\mathfrak{X}, \mathbf{l})$. This number is known as the width of $\mathfrak{X}$ in the direction $\mathbf{l}$.

The upper and lower bounds of $w(\mathfrak{X}, \mathbf{l})$ are both attained, denote them respectively by w^* and w_* and let $\mathbf{l}^*$ and $\mathbf{l}_*$ be two directions such that

$$w(\mathfrak{X}, \mathbf{l}^*) = w^*, \quad w(\mathfrak{X}, \mathbf{l}_*) = w_*. \tag{4·5·1}$$

On each of the two hyperplanes perpendicular to $\mathbf{l}^*$ there is a point of $\mathfrak{X}$. Thus if we use $D(\mathfrak{X})$ for the diameter of $\mathfrak{X}$ as before we have

$$w^* \leqslant D(\mathfrak{X}).$$

But there are two points $\mathbf{x}_1, \mathbf{x}_2$ of $\mathfrak{X}$ such that

$$|\mathbf{x}_1 - \mathbf{x}_2| = D(\mathfrak{X}).$$

Consider the two hyperplanes that are perpendicular to $\mathbf{x}_1\mathbf{x}_2$ and pass one each through $\mathbf{x}_1$ and $\mathbf{x}_2$. These hyperplanes are support hyperplanes of $\mathfrak{X}$, for otherwise we could find two points of $\mathfrak{X}$ at a distance apart greater than $D(\mathfrak{X})$. Thus it follows that

$$w^* \geqslant D(\mathfrak{X}).$$

Hence finally the maximum width of $\mathfrak{X}$ is equal to its diameter.

We shall denote the lower bound w_* by $d(\mathfrak{X})$ and refer to it as the minimal width of $\mathfrak{X}$.

A useful property of the two support hyperplanes perpendicular to $\mathbf{l}_*$, where $w(\mathfrak{X}, \mathbf{l}_*) = d(\mathfrak{X})$ is the following. Denote the two hyperplanes by $\mathfrak{P}_1$ and $\mathfrak{P}_2$. Then if $\mathfrak{P}_1 \cap \mathfrak{X}$ is projected in the direction $\mathbf{l}_*$ onto $\mathfrak{P}_2$, its projection intersects $\mathfrak{P}_2 \cap \mathfrak{X}$. If this were not the case there would be an $n-2$ space in $\mathfrak{P}_2$ separating the projection of $\mathfrak{P}_1 \cap \mathfrak{X}$ from $\mathfrak{P}_2 \cap \mathfrak{X}$ by Theorem 7. Denote this $(n-2)$-dimensional space by $\mathfrak{Q}_2$. The hyperplane spanned by $\mathbf{l}_*$ and $\mathfrak{Q}_2$ intersects $\mathfrak{P}_1$ in say $\mathfrak{Q}_1$. We can rotate $\mathfrak{P}_1$ about $\mathfrak{Q}_1$ and $\mathfrak{P}_2$ about $\mathfrak{Q}_2$ so that in their new positions $\mathfrak{P}_1$ and $\mathfrak{P}_2$ are parallel and do not meet $\mathfrak{X}$. Since their distance apart is less than $d(\mathfrak{X})$ we have a contradiction with the definition of $d(\mathfrak{X})$.

A set $\mathfrak{X}$ for which $D(\mathfrak{X}) = d(\mathfrak{X})$ is said to be of constant width.

(iii) *The centre of a convex set* $\mathfrak{X}$

If there is a point $\mathbf{p}$ such that $\mathfrak{X}$ coincides with its reflexion in $\mathbf{p}$, then $\mathbf{p}$ is said to be the centre of $\mathfrak{X}$ and $\mathfrak{X}$ is said to be central. The properties of central convex sets are usually easier to establish than those of non-central convex sets.

(iv) *The centroid of a convex set* $\mathfrak{X}$

This is the centroid of a uniform mass distributed throughout $\mathfrak{X}$.

Exercises 4·5

1. Show that $d(\mathfrak{X})$, $D(\mathfrak{X})$, $R(\mathfrak{X})$, $r(\mathfrak{X})$ depend continuously on $\mathfrak{X}$, and that $w(\mathfrak{X}, \mathbf{l})$ depends continuously on $\mathfrak{X}$ and $\mathbf{l}$.

2. $\mathfrak{X}$ is a convex set and $\mathfrak{K}$ is the union of all the inspheres of $\mathfrak{X}$. Show that $\mathfrak{K}$ is convex.

3. Show that a bounded convex set can have at most one centre, but that an unbounded convex set can have infinitely many. Show that in this second case the centres form a linear subspace of R^n.

CHAPTER 5

TRANSFORMATIONS AND COMBINATIONS OF CONVEX SETS

The volume of a convex set $\mathfrak{X}$, denoted by $V(\mathfrak{X})$, was defined in Chapter 4 as its n-dimensional Lebesgue measure. No mention was made then of the surface area of a convex set. The definition and some of the basic properties of surface area are given in this chapter.

It would be possible to define the surface area of $\mathfrak{X}$ as the $(n-1)$-dimensional Hausdorff measure of the frontier of $\mathfrak{X}$. Such a definition, besides depending on concepts which do not arise naturally in the theory of convex sets, would be difficult to handle, and it would not be easy to establish certain properties that we shall require.

Instead we give an account of Minkowski's definition of surface area. This definition is embedded in the theory of linear arrays and of mixed volumes, a theory which is of great importance in the elucidation of geometrical relations between convex sets. A brief account of this theory is given and a proof of the Brünn-Minkowski theorem is given in § 5·5. An alternative proof of this theorem is that due to Blaschke and depends on Steiner symmetrization which is defined and developed in § 5·4. The first proof of the Brünn-Minkowski theorem is also valid for non-convex sets. The theory of mixed volumes, on the other hand, cannot be extended to non-convex sets; the idea of the mixed volume depends essentially on the convexity of the sets concerned. Inequalities between mixed volumes due to Minkowski, Fenchel and Alexandroff are also given in § 5·5. The chapter also contains a short account of central symmetrization.

1. Linear and concave† arrays of convex sets

Let $\mathfrak{X}_1, \ldots, \mathfrak{X}_r$ be r given closed bounded convex sets and let the point $\mathbf{p}$ be taken as origin. Denote by $\mathbf{x}_i$ the vector

† These arrays are also sometimes called convex.

from $\mathbf{p}$ to a variable point in $\mathfrak{X}_i$ and define the point or vector $\mathbf{x}$ by

$$\mathbf{x} = \lambda_1 \mathbf{x}_1 + \lambda_2 \mathbf{x}_2 + \ldots + \lambda_r \mathbf{x}_r,$$

where $\lambda_1, \lambda_2, \ldots, \lambda_r$ are fixed non-negative numbers. As $\mathbf{x}_1, \mathbf{x}_2, \ldots, \mathbf{x}_r$ vary independently in $\mathfrak{X}_1, \mathfrak{X}_2, \ldots, \mathfrak{X}_r$, the point $\mathbf{x}$ describes a set $\mathfrak{X}$ whose relationship to $\mathfrak{X}_1, \ldots, \mathfrak{X}_r$ is denoted by

$$\mathfrak{X} = \lambda_1 \mathfrak{X}_1 + \lambda_2 \mathfrak{X}_2 + \ldots + \lambda_r \mathfrak{X}_r.$$

The addition defined by this process is called the vector addition of the sets concerned and is both associative and commutative. If all the sets $\mathfrak{X}_i$ are closed, bounded and convex, then it may be shown that $\mathfrak{X}$ also is closed, bounded and convex. (The convexity of $\mathfrak{X}$ follows from an argument similar to that used at the end of § 2, Chapter 1).

The set $\mathfrak{X}$ is defined by means of vectors $\mathbf{x}_i$, and the value of these vectors depends upon the particular point chosen as origin. Thus $\mathfrak{X}$ itself may change if we choose another point as the origin. If we select a new point $\mathbf{p}'$ as origin (where $\mathbf{p}'$ denotes the vector from the old origin to the new), then the set $\mathfrak{X}$ is replaced by the set obtained from $\mathfrak{X}$ by subjecting it to a translation $\mathbf{p}'(1 - \lambda_1 - \ldots - \lambda_r)$. Thus if $\lambda_1 + \ldots + \lambda_r$ is equal to unity, $\mathfrak{X}$ is independent of the choice of origin, and in any case the different sets $\mathfrak{X}$ which are obtained by using different points as origin are all congruent to one another.

Let the support function of $\mathfrak{X}_i$ be $H_i(\mathbf{u})$. Then the support function of $\mathfrak{X}$ is

$$H(\mathbf{u}) = \sup_{\mathbf{x} \in \mathfrak{X}} (\mathbf{x} \,.\, \mathbf{u}) = \lambda_1 H_1(\mathbf{u}) + \ldots + \lambda_r H_r(\mathbf{u}). \qquad (5{\cdot}1{\cdot}1)$$

We regard this *linear array* of convex sets as a mapping of a convex subset of real r-dimensional Euclidean space, R^r, onto a class of convex subsets of R^n defined by

$$(\lambda_1, \ldots, \lambda_r) \to \lambda_1 \mathfrak{X}_1 + \ldots + \lambda_r \mathfrak{X}_r.$$

More generally we can make correspond to each point

$$\mathbf{d} = (d_1, \ldots, d_r)$$

of a convex subset of R^r a closed bounded convex subset of R^n, say $\mathfrak{X}(\mathbf{d})$, such that for any θ, $0 \leqslant \theta \leqslant 1$,

$$(1-\theta)\,\mathfrak{X}(\mathbf{d}_0) + \theta \mathfrak{X}(\mathbf{d}_1) \subset \mathfrak{X}((1-\theta)\,\mathbf{d}_0 + \theta \mathbf{d}_1). \qquad (5{\cdot}1{\cdot}2)$$

Such a class of convex sets is said to form a *concave array*. Every linear array is a concave array but the converse is not true as we shall soon see.

Consider a concave array defined over a segment of R^1, say $\mathfrak{X}(\lambda)$, where λ is a real number and $0 \leqslant \lambda \leqslant 1$. Suppose that the sets $\mathfrak{X}(\lambda)$ lie in R^n and that coordinates are taken in a space R^{n+1} such that R^n is the space $x_{n+1}=0$. Define $\mathfrak{Y}(\lambda)$ to be the set of points $(y_1, \ldots, y_{n+1})$, where $y_{n+1}=\lambda$ and $(y_1, \ldots, y_n, 0) \in \mathfrak{X}(\lambda)$. Let $\mathfrak{Y}$ be the union of all the sets $\mathfrak{Y}(\lambda)$. It may be verified that the convexity of the sets $\mathfrak{Y}(\lambda)$ together with (5·1·2) implies that $\mathfrak{Y}$ is a convex set. Conversely, if we take a closed bounded convex set $\mathfrak{Y}$ in R^{n+1}, the projections of the sections of $\mathfrak{Y}$ by n-dimensional linear manifolds parallel to $x_{n+1}=0$ onto $x_{n+1}=0$ give rise to a concave array. Further, this array is linear if and only if $\mathfrak{Y}$ is the convex cover of its two extreme sections, i.e. the sections by the n-spaces $x_{n+1}=\lambda$, where λ has the largest and least values such that $x_{n+1}=\lambda$ has a non-empty intersection with $\mathfrak{Y}$. Since there are convex sets $\mathfrak{Y}$ not of this form it follows that there are concave non-linear arrays.

The following properties of linear and concave arrays also hold:

THEOREM 38. (*a*) *If every set of a linear or concave array is projected onto a linear subspace of* R^n, *the projected sets again form a linear or concave array respectively.*

(*b*) *If* $\mathfrak{X}=\lambda_1\mathfrak{X}_1+\ldots+\lambda_r\mathfrak{X}_r$ *and the support hyperplanes of* $\mathfrak{X}_i$, $\mathfrak{X}$ *perpendicular to* $\mathbf{u}$ *and similarly situated with respect to* $\mathfrak{X}_i$, $\mathfrak{X}$ *are* $\mathfrak{P}_i$, $\mathfrak{P}$ *respectively, then*

$$\mathfrak{X} \cap \mathfrak{P} = \lambda_1\mathfrak{X}_1 \cap \mathfrak{P}_1 + \lambda_2\mathfrak{X}_2 \cap \mathfrak{P}_2 + \ldots + \lambda_r\mathfrak{X}_r \cap \mathfrak{P}_r.$$

The proof of (*a*) is immediate, since any vector can be expressed as the union of two components, one lying in the linear subspace and one perpendicular to this subspace. Any linear relation which holds amongst the vectors continues to hold for these components.

To prove (*b*) we differentiate (5·1·1) in the direction $\mathbf{v}$. Then

$$\delta H(\mathbf{u};\, \mathbf{v}) = \lambda_1 \delta H_1(\mathbf{u};\, \mathbf{v}) + \ldots + \lambda_r \delta H_r(\mathbf{u};\, \mathbf{v}). \qquad (5\cdot1\cdot3)$$

But $\delta H_i(\mathbf{u};\, \mathbf{v})$ is the support function of $\mathfrak{X}_i \cap \mathfrak{P}_i$, and the equation (5·1·3) implies the identity stated in (*b*).

Remarks. (i) If $\mathfrak{X}_i$ is given a translation $\mathbf{v}_i, i=1,2,\dots,r$, then $\mathfrak{X}=\lambda_1\mathfrak{X}_1+\dots+\lambda_r\mathfrak{X}_r$ undergoes a translation $\lambda_1\mathbf{v}_1+\dots+\lambda_r\mathbf{v}_r$. Thus if we are considering some property of $\mathfrak{X}$ that is invariant under translations, then we can give to $\mathfrak{X}_1,\dots,\mathfrak{X}_r$ such positions as we choose provided their orientation is not altered. In particular, we can suppose that they all contain a given point.

(ii) If the same affine transformation, say $\mathbf{y}\to\mathbf{y}'$, where $\mathbf{y}'=\mathrm{A}\mathbf{y}+\mathbf{b}$ is applied to each of the $\mathfrak{X}_i$, then $\mathfrak{X}$ is subjected to the affine transformation $\mathbf{x}\to\mathbf{x}'$, where

$$\mathbf{x}'=\lambda_1\mathbf{x}_1'+\dots+\lambda_r\mathbf{x}_r'=\mathrm{A}\mathbf{x}+(\lambda_1+\dots+\lambda_r)\,\mathbf{b}.$$

Thus if each $\mathfrak{X}_i$ is subjected to the same volume-preserving affine transformation the set $\mathfrak{X}$ is also subjected to a volume-preserving affine transformation.

Exercises 5·1

1. $\mathbf{x}$ is an extreme point of the convex set $\mathfrak{X}$ defined by

$$\mathfrak{X}=\lambda_1\mathfrak{X}_1+\lambda_2\mathfrak{X}_2+\dots+\lambda_r\mathfrak{X}_r \quad (\lambda_i>0, 1\leqslant i\leqslant r).$$

Show that if $\mathbf{x}=\lambda_1\mathbf{x}_1+\lambda_2\mathbf{x}_2+\dots+\lambda_r\mathbf{x}_r$, where $\mathbf{x}_i\in\mathfrak{X}_i$, then each $\mathbf{x}_i$ is an extreme point of $\mathfrak{X}_i$.

2. $\mathfrak{X}$ is a convex set and $\mathfrak{S}$ a sphere whose centre is the origin. A support hyperplane $\mathfrak{P}$ of $\mathfrak{X}$ cuts the set $\mathfrak{X}+\mathfrak{S}$ into two sets $\mathfrak{X}_1,\mathfrak{X}_2$, where $\mathfrak{X}_2\supset\mathfrak{X}$. Show that the reflexion of $\mathfrak{X}_1$ in $\mathfrak{P}$ is contained in $\mathfrak{X}_2$. Give an example to show that this result need not be true when $\mathfrak{S}$ is replaced by an arbitrary convex set.

3. $\mathfrak{X}$ is a convex set and $\mathfrak{S}$ a sphere whose centre is the origin. Show that the centroid of $\mathfrak{X}+\mathfrak{S}$ is contained in $\mathfrak{X}$. Give an example to show that this need not be true when $\mathfrak{S}$ is replaced by an arbitrary convex set.

4. $\mathfrak{X}_1$ and $\mathfrak{X}_2$ are two given convex sets. Show that there are two parallel hyperplanes $\mathfrak{P}_1$ and $\mathfrak{P}_2$ such that $\mathfrak{P}_1$ is a regular support hyperplane of $\mathfrak{X}_1$, and $\mathfrak{P}_2$ is a regular support hyperplane of $\mathfrak{X}_2$. (*Hint.* Consider the points of $\frac{1}{2}(\mathfrak{X}_1+\mathfrak{X}_2)$ which lie on regular support hyperplanes.)

2. Mixed volumes

A concept of considerable importance in the theory of convex sets is that of mixed volumes. We shall define these numbers in the first place for linear arrays of polytopes.

Suppose that $\mathfrak{X}_1, \mathfrak{X}_2, \ldots, \mathfrak{X}_r$ are r convex polytopes and we form the linear array

$$\mathfrak{X} = \lambda_1 \mathfrak{X}_1 + \ldots + \lambda_r \mathfrak{X}_r. \tag{5·2·1}$$

Since every extreme point $\mathbf{x}$ of $\mathfrak{X}$ is a linear combination of extreme points $\mathbf{x}_i$ of $\mathfrak{X}_i$ of the form

$$\mathbf{x} = \lambda_1 \mathbf{x}_1 + \ldots + \lambda_r \mathbf{x}_r,$$

it follows that there are at most a finite number of extreme points of $\mathfrak{X}$ and that $\mathfrak{X}$ itself is a convex polytope. We wish to determine the volume of $\mathfrak{X}$ given by (5·2·1) in terms of $\lambda_1, \ldots, \lambda_r$.

Since $\mathfrak{X}$ is a polytope it has a finite number of $(n-1)$-dimensional faces which we may suppose to lie in the hyperplanes $\mathfrak{P}_1, \mathfrak{P}_2, \ldots, \mathfrak{P}_k$. Then, by (4·4·4),

$$V(\mathfrak{X}) = \frac{1}{n} \sum_{j=1}^{k} H(\mathbf{u}_j) V_{n-1}(\mathfrak{P}_j \cap \mathfrak{X}), \tag{5·2·2}$$

where $\mathbf{u}_j$ is the unit normal to $\mathfrak{P}_j$ in the sense outward from $\mathfrak{X}$. For the function $H(\mathbf{u}_j)$ we have the expression

$$H(\mathbf{u}_j) = \sum_{i=1}^{r} \lambda_i H_i(\mathbf{u}_j), \tag{5·2·3}$$

and for the function $V_{n-1}(\mathfrak{P}_j \cap \mathfrak{X})$, we have, applying (5·2·2) in the $(n-1)$-dimensional linear manifold $\mathfrak{P}_j$, an expression of the form

$$V_{n-1}(\mathfrak{P}_j \cap \mathfrak{X}) = \frac{1}{n-1} \sum_{h=1}^{k_j} \delta H(\mathbf{u}_j;\, \mathbf{v}_h) V_{n-2}(\mathfrak{P}_{j,h} \cap \mathfrak{X}), \tag{5·2·4}$$

where $\mathfrak{P}_{j,h}$, $h = 1, 2, \ldots, k_j$, denote the $(n-2)$-dimensional linear manifolds that lie in $\mathfrak{P}_j$ and meet $\mathfrak{X}$ in $(n-2)$-dimensional sets and $\mathbf{v}_h$ is the unit vector in $\mathfrak{P}_j$ normal to $\mathfrak{P}_{j,h}$ and in the sense outwards from $\mathfrak{X} \cap \mathfrak{P}_j$. Thus substituting from (5·2·3) and (5·2·4) in (5·2·2) we obtain

$$V(\mathfrak{X}) = \frac{1}{n(n-1)} \sum_{j=1}^{k} \sum_{h=1}^{k_j} \sum_{i=1}^{r} \lambda_i H_i(\mathbf{u}_j)\, \delta H(\mathbf{u}_j;\, \mathbf{v}_h) V_{n-2}(\mathfrak{P}_{j,h} \cap \mathfrak{X}). \tag{5·2·5}$$

Next use

$$\delta H(\mathbf{u}_j;\, \mathbf{v}_h) = \sum_{i=1}^{r} \lambda_i \delta H_i(\mathbf{u}_j;\, \mathbf{v}_h) \tag{5·2·6}$$

and

$$V_{n-2}(\mathfrak{P}_{j,h} \cap \mathfrak{X}) = \frac{1}{n-2} \sum_{g=1}^{k_{j,h}} \delta^2 H(\mathbf{u}_j;\, \mathbf{v}_h;\, \mathbf{w}_g) V_{n-3}(\mathfrak{P}_{j,h,g} \cap \mathfrak{X}), \tag{5·2·7}$$

where $\mathfrak{P}_{j,h,g}$, $g=1,2,\ldots,k_{j,h}$, denote the $(n-3)$-dimensional linear manifolds that lie in $\mathfrak{P}_{j,h}$ and meet $\mathfrak{X}$ in $(n-3)$-dimensional sets and $\mathbf{w}_g$ is the unit vector in $\mathfrak{P}_{j,h}$ normal to $\mathfrak{P}_{j,h,g}$ and directed outwards from $\mathfrak{X} \cap \mathfrak{P}_{j,h}$. Substitute these values from (5·2·6) and (5·2·7) in (5·2·5). The process is repeated a finite number of times and we obtain an expression of the form

$$V(\mathfrak{X}) = \frac{1}{n!} \sum_{j=1}^{k} \sum_{h=1}^{k_j} \sum_{g=1}^{k_{j,h}} \cdots \sum_{i_1=1}^{r} \sum_{i_2=1}^{r} \cdots \sum_{i_n=1}^{r} \lambda_{i_1} \lambda_{i_2} \ldots \lambda_{i_n} H_{i_1}(\mathbf{u}_j)\, \delta H_{i_2}(\mathbf{u}_j; \mathbf{v}_h) \ldots. \tag{5·2·8}$$

The important properties of (5·2·8) are summarized in the following theorem:

THEOREM 39. *If $\mathfrak{X} = \lambda_1 \mathfrak{X}_1 + \ldots + \lambda_r \mathfrak{X}_r$ is a linear array of convex polytopes then the volume of $\mathfrak{X}$ is a homogeneous n-th degree polynomial in the variables $\lambda_1, \ldots, \lambda_r$,*

$$V(\mathfrak{X}) = \sum_{i_1=1}^{r} \sum_{i_2=1}^{r} \cdots \sum_{i_n=1}^{r} V(i_1, i_2, \ldots, i_n)\, \lambda_{i_1} \ldots \lambda_{i_n}, \tag{5·2·9}$$

where the coefficients $V(i_1, i_2, \ldots, i_n)$ are chosen to be invariant under permutations of their arguments.

DEFINITION. *The coefficients $V(i_1, \ldots, i_n)$ are defined to be the mixed volumes of the array.*

When it is desirable to indicate the sets used in defining the array explicitly we use $V(\mathfrak{X}_{i_1}, \ldots, \mathfrak{X}_{i_n})$ in place of $V(i_1, \ldots, i_n)$. The number of arguments of a mixed volume is always equal to the dimension of the containing space. A mixed volume obtained from two sets $\mathfrak{Y}_1, \mathfrak{Y}_2$, say $V(\mathfrak{X}_1, \ldots, \mathfrak{X}_n)$, where $\mathfrak{X}_i = \mathfrak{Y}_1$, $1 \leqslant i \leqslant s$; $\mathfrak{X}_i = \mathfrak{Y}_2$, $s+1 \leqslant i \leqslant n$, will be denoted by $V(\mathfrak{Y}_1, s; \mathfrak{Y}_2, n-s)$.

COROLLARY 1. *The mixed volumes are all non-negative.*

The formula (5·2·2) is independent of the position of the origin; thus we may suppose all the $H(\mathbf{u}_j)$ are non-negative by taking as origin a point of $\mathfrak{X}$. Similarly in (5·2·4) we can take as origin a point of $\mathfrak{P}_j \cap \mathfrak{X}$ (so that we have different origins corresponding to different faces of $\mathfrak{X}$) and then each function $\delta H(\mathbf{u}_j; \mathbf{v}_h)$ is non-negative. Proceeding in this way we see that all the factors of the coefficients of (5·2·8) are non-negative and this establishes the corollary.

COROLLARY 2. *If the sets $\mathfrak{X}_1, \ldots, \mathfrak{X}_r$ are all contained in a certain convex set $\mathfrak{Y}$ of volume K then $V(i_1, \ldots, i_n) \leqslant K . r^n$.*

For in (5·2·9) take $\lambda_1 = \lambda_2 = \ldots = \lambda_r = 1/r$. The set $\mathfrak{X}$ is then contained in $\mathfrak{Y}$. Thus, since all the terms of (5·2·9) are non-negative,

$$K \geqslant V(\mathfrak{X}) \geqslant V(i_1, i_2, \ldots, i_n)\, r^{-n},$$

and the corollary is proved.

We next extend the concept of mixed volume by considering the case of r convex sets $\mathfrak{X}_1, \ldots, \mathfrak{X}_r$ which are not necessarily polytopes. We can select r sequences of polytopes

$$\{\mathfrak{Q}_{i,j}\} \quad (i = 1, \ldots, r,\ j = 1, 2, \ldots),$$

such that $\mathfrak{Q}_{i,j}$ converges to $\mathfrak{X}_i$ as j tends to infinity. Then

$$\mathfrak{Q}_j = \lambda_1 \mathfrak{Q}_{1,j} + \ldots + \lambda_r \mathfrak{Q}_{r,j}$$

converges to

$$\mathfrak{X} = \lambda_1 \mathfrak{X}_1 + \ldots + \lambda_r \mathfrak{X}_r$$

as j tends to infinity. Hence $V(\mathfrak{Q}_j)$ tends to $V(\mathfrak{X})$. The mixed volumes $V(\mathfrak{Q}_{i_1,j}, \mathfrak{Q}_{i_2,j}, \ldots, \mathfrak{Q}_{i_n,j})$ are all bounded as j varies. We may select a subsequence of integers j_k such that each mixed volume converges to a limit as k tends to infinity. Thus it follows that $V(\mathfrak{X})$ is a polynomial in $\lambda_1, \ldots, \lambda_r$ which is homogeneous of the nth degree. This polynomial is written so that its coefficients are unchanged by a permutation of their arguments. It is the same polynomial whatever sequences $\mathfrak{Q}_{i,j}$ are used and its coefficients are called the mixed volumes of $\mathfrak{X}_1, \ldots, \mathfrak{X}_r$. The corollaries of Theorem 39 hold for these mixed volumes also.

An important property of $V(\mathfrak{X}_{i_1}, \ldots, \mathfrak{X}_{i_n})$ is that it increases if we enlarge any of the sets $\mathfrak{X}_i$. We prove first Theorems 40 and 41.

THEOREM 40. *If $H(\mathbf{u})$ is the support function of the convex set $\mathfrak{X}$ and if the polytope $\mathfrak{Q}$ has faces $\mathfrak{Q}_1, \ldots, \mathfrak{Q}_k$ such that $\mathbf{u}_i$ is the unit vector normal to the hyperplane containing $\mathfrak{Q}_i$ and in the sense outwards from $\mathfrak{X}$, then*

$$V(\mathfrak{X}, 1; \mathfrak{Q}, n-1) = \frac{1}{n} \sum_{i=1}^{k} H(\mathbf{u}_i)\, V_{n-1}(\mathfrak{Q}_i). \qquad (5·2·10)$$

We have from the definition of a mixed volume the relation

$$V(\mathfrak{X}, 1; \mathfrak{Q}, n-1) = \frac{1}{n} \lim_{\delta \to 0} \frac{V(\mathfrak{Q} + \delta\mathfrak{X}) - V(\mathfrak{Q})}{\delta}. \qquad (5·2·11)$$

Let $\mathbf{x}_i$ be a point of $\mathfrak{X}$ such that $\mathbf{x}_i . \mathbf{u}_i = H(\mathbf{u}_i)$. As the point $\mathbf{q}$ describes the face $\mathfrak{Q}_i$ of $\mathfrak{Q}$, $\mathbf{q} + \mathbf{x}_i$ describes a congruent set in a parallel hyperplane at a perpendicular distance from $\mathfrak{Q}_i$ of $H(\mathbf{u}_i)$. Thus

$$V(\mathfrak{Q} + \delta\mathfrak{X}) - V(\mathfrak{Q}) \geqslant \sum_{i=1}^{k} \delta . H(\mathbf{u}_i) V_{n-1}(\mathfrak{Q}_i). \qquad (5\cdot2\cdot12)$$

On the other hand, the part of $\mathfrak{Q} + \delta\mathfrak{X}$ that belongs neither to $\mathfrak{Q}$ nor to one of the k cylinders whose volumes form the sum on the right-hand side of (5·2·12) is contained in $\mathfrak{U}(\mathfrak{R}, \delta_1)$, where $\mathfrak{R}$ is the union of all the $(n-2)$-dimensional faces of $\mathfrak{Q}$, and $\delta_1 = \delta D(\mathfrak{X})$. $\mathfrak{R}$ is of finite $(n-2)$-dimensional measure and by Theorem 36

$$V(\mathfrak{U}(\mathfrak{R}, \delta_1)) = O(\delta^2). \qquad (5\cdot2\cdot13)$$

(5·2·13) together with (5·2·12) and (5·2·11) implies the relation (5·2·10).

The second preliminary result is

THEOREM 41. *If $\mathfrak{X}_1, \mathfrak{X}_2, \ldots, \mathfrak{X}_n$ are given convex sets and*

$$\mathfrak{K} = \lambda_2 \mathfrak{X}_2 + \ldots + \lambda_n \mathfrak{X}_n,$$

then $$V(\mathfrak{X}_1, 1; \mathfrak{K}, n-1) = \Sigma V(\mathfrak{X}_1, \mathfrak{X}_{i_2}, \ldots, \mathfrak{X}_{i_n}) \lambda_{i_2} \ldots \lambda_{i_n},$$

where in the summation each suffix i_j varies independently between 2 and n.

Apply formula (5·2·9) to obtain two expressions for $V(\mathfrak{Z})$, where

$$\mathfrak{Z} = \lambda_1 \mathfrak{X}_1 + \lambda_2 \mathfrak{X}_2 + \ldots + \lambda_n \mathfrak{X}_n = \lambda_1 \mathfrak{X}_1 + \mathfrak{K}.$$

Then

$$V(\mathfrak{Z}) = \Sigma V(\mathfrak{X}_{i_1}, \ldots, \mathfrak{X}_{i_n}) \lambda_{i_1}, \ldots, \lambda_{i_n} = \Sigma \tbinom{n}{\nu} V(\mathfrak{X}_1, \nu; \mathfrak{K}, n-\nu) . \lambda_1^{\nu}.$$

Equate coefficients of λ_1, and we obtain the required identity.

We can now prove that $V(\mathfrak{X}_1, \ldots, \mathfrak{X}_n)$ is an increasing function of its arguments.

THEOREM 42. *If $\mathfrak{X}_i$ and $\mathfrak{X}_i'$ are such that $\mathfrak{X}_i \subset \mathfrak{X}_i'$, $i = 1, 2, \ldots, n$, then $V(\mathfrak{X}_1, \ldots, \mathfrak{X}_n) \leqslant V(\mathfrak{X}_1', \ldots, \mathfrak{X}_n')$.*

It is sufficient to prove that if $\mathfrak{X} \subset \mathfrak{Y}$, then

$$V(\mathfrak{X}, \mathfrak{X}_2, \ldots, \mathfrak{X}_n) \leqslant V(\mathfrak{Y}, \mathfrak{X}_2, \ldots, \mathfrak{X}_n). \qquad (5\cdot2\cdot14)$$

Further it is sufficient to consider the case when the sets $\mathfrak{X}_2, \ldots, \mathfrak{X}_n$ are polytopes, for the result in the general case can then be deduced by an approximation argument.

We may suppose that the origin is contained in $\mathfrak{X}$ so that if $H_X(\mathbf{u})$ and $H_Y(\mathbf{u})$ are the support functions of $\mathfrak{X}$ and $\mathfrak{Y}$ respectively, we have

$$0 \leqslant H_X(\mathbf{u}) \leqslant H_Y(\mathbf{u}). \tag{5·2·15}$$

Let $\mathfrak{Q}$ be the polytope defined by

$$\mathfrak{Q} = \lambda_2 \mathfrak{X}_2 + \ldots + \lambda_n \mathfrak{X}_n.$$

Then by Theorem 40

$$V(\mathfrak{X}, 1; \mathfrak{Q}, n-1) = \frac{1}{n} \Sigma H_X(\mathbf{u}_i) V_{n-1}(\lambda_2 \mathfrak{Q}_i^{(2)} + \ldots + \lambda_n \mathfrak{Q}_i^{(n)}), \tag{5·2·16}$$

where $\mathbf{u}_i$ is the outward unit vector normal to the ith face $\mathfrak{Q}_i$ of $\mathfrak{Q}$, $\mathfrak{Q}_i^{(r)}$ is the at most $(n-1)$-dimensional face of $\mathfrak{X}_r$ perpendicular to $\mathbf{u}_i$ and we have used the fact that $\mathfrak{Q}_i$ is $\lambda_2 \mathfrak{Q}_i^{(2)} + \ldots + \lambda_n \mathfrak{Q}_i^{(n)}$. (The normals to the $(n-1)$-dimensional faces of $\mathfrak{Q}$ are parallel to one or other of a fixed set of vectors which are independent of the $\lambda_2, \ldots, \lambda_n$ and we suppose that the summation is over all such normals.) By Theorem 41 the left-hand side of (5·2·16) is a homogeneous polynomial in $\lambda_2, \ldots, \lambda_n$ of degree $n-1$, and by theorem 39 applied in an $(n-1)$-dimensional linear manifold the same is true of the right-hand side of (5·2·16). Equate coefficients of $\lambda_2 \lambda_3 \ldots \lambda_n$, then we obtain

$$V(\mathfrak{X}, \mathfrak{X}_2, \ldots, \mathfrak{X}_n) = \frac{1}{n} \Sigma H_X(\mathbf{u}_i)\, v(\mathfrak{Q}_i^{(2)}, \ldots, \mathfrak{Q}_i^{(n)}), \tag{5·2·17}$$

where v is used to denote an $(n-1)$-dimensional mixed volume. A similar formula holds for $V(\mathfrak{Y}, \mathfrak{X}_2, \ldots, \mathfrak{X}_n)$ and from (5·2·15) and the fact that mixed volumes are always non-negative we deduce the correctness of (5·2·14).

Thus the theorem is proved.

COROLLARY. *If $\mathfrak{X}_1, \ldots, \mathfrak{X}_r$ are closed bounded convex sets then each mixed volume $V(\mathfrak{X}_{i_1}, \ldots, \mathfrak{X}_{i_n})$ varies continuously with $\mathfrak{X}_1, \ldots, \mathfrak{X}_r$.*

We have to show that if $\{\mathfrak{Y}_{i,j}\}$, $i = 1, 2, \ldots, r$, $j = 1, 2, \ldots$, are r sequences of convex sets such that $\mathfrak{Y}_{i,j} \to \mathfrak{X}_i$ as $j \to \infty$, then

$$\lim_{j \to \infty} V(\mathfrak{Y}_{i_1,j}, \mathfrak{Y}_{i_2,j}, \ldots, \mathfrak{Y}_{i_n,j}) = V(\mathfrak{X}_{i_1}, \ldots, \mathfrak{X}_{i_n}).$$

If

$$\mathfrak{Y}_{i,j} \supset \mathfrak{X}_i \quad (i = 1, 2, \ldots, r),$$

then

$$V(\mathfrak{Y}_{i_1,j}, \ldots, \mathfrak{Y}_{i_n,j}) \geqslant V(\mathfrak{X}_{i_1}, \ldots, \mathfrak{X}_{i_n}).$$

As j tends to infinity

$$\mathfrak{Y} = \lambda_1 \mathfrak{Y}_{1,j} + \ldots + \lambda_r \mathfrak{Y}_{r,j} \to \mathfrak{X} = \lambda_1 \mathfrak{X}_1 + \ldots + \lambda_r \mathfrak{X}_r,$$

and thus

$$V(\mathfrak{Y}) = \Sigma V(\mathfrak{Y}_{i_1,j}, \ldots, \mathfrak{Y}_{i_n,j}) \lambda_{i_1} \ldots \lambda_{i_n}$$

$$\to V(\mathfrak{X}) = \Sigma V(\mathfrak{X}_{i_1}, \ldots, \mathfrak{X}_{i_n}) \lambda_{i_1} \ldots \lambda_{i_n}. \quad (5\cdot2\cdot18)$$

This is possible only if

$$V(\mathfrak{Y}_{i_1,j}, \ldots, \mathfrak{Y}_{i_n,j}) \to V(\mathfrak{X}_{i_1}, \ldots, \mathfrak{X}_{i_n}).$$

In particular

$$\lim_{\delta \to 0} V(\overline{\mathfrak{U}(\mathfrak{X}_{i_1}, \delta)}, \ldots, \overline{\mathfrak{U}(\mathfrak{X}_{i_n}, \delta)}) = V(\mathfrak{X}_{i_1}, \ldots, \mathfrak{X}_{i_n})$$

Now $\mathfrak{Y}_{i,j} \to \mathfrak{X}_i$ as $j \to \infty$ implies that given $\delta > 0$ there is an integer j such that $\mathfrak{Y}_{i,j} \subset \mathfrak{U}(\mathfrak{X}_i, \delta)$ and therefore

$$\overline{\lim_{j \to \infty}} V(\mathfrak{Y}_{i_1,j}, \ldots, \mathfrak{Y}_{i_n,j}) \leqslant V(\mathfrak{X}_{i_1}, \ldots, \mathfrak{X}_{i_n}).$$

But from (5·2·18) this is possible only if

$$\lim_{j \to \infty} V(\mathfrak{Y}_{i_1,j}, \ldots, \mathfrak{Y}_{i_n,j}) = V(\mathfrak{X}_{i_1}, \ldots, \mathfrak{X}_{i_n}).$$

3. Surface area

DEFINITION. *The surface area of a bounded closed convex set* $\mathfrak{X}$ *is* $A(\mathfrak{X})$ *defined by*

$$A(\mathfrak{X}) = \lim_{\delta \to 0+} \frac{V(\mathfrak{U}(\mathfrak{X}, \delta)) - V(\mathfrak{X})}{\delta}. \quad (5\cdot3\cdot1)$$

The fact that this limit exists follows from the fact that $A(\mathfrak{X})$ is a mixed volume. We have

$$\mathfrak{U}(\mathfrak{X}, \delta) = \mathfrak{X} + \delta\mathfrak{S},$$

where $\mathfrak{S}$ is the unit sphere whose centre is the origin, and thus, from Theorem 39,

$$V(\mathfrak{U}(\mathfrak{X}, \delta)) = V(\mathfrak{X}) + nV(\mathfrak{X}, n-1;\ \mathfrak{S}, 1)\,\delta + O(\delta^2). \quad (5\cdot3\cdot2)$$

Thus $A(\mathfrak{X})$ is equal to $nV(\mathfrak{X}, n-1;\ \mathfrak{S}, 1)$ and the limit (5·3·1) exists. From the corresponding properties for mixed volumes we see that

(*a*) $A(\mathfrak{X})$ depends continuously on $\mathfrak{X}$, (*b*) if $\mathfrak{Y} \supset \mathfrak{X}$, $A(\mathfrak{Y}) \geqslant A(\mathfrak{X})$,

(*c*) if $\mathfrak{Q}$ is a polytope $A(\mathfrak{Q})$ is equal to the sum of the $(n-1)$-dimensional measures of its faces.

We can strengthen the inequality (*b*) by remarking that if $\mathfrak{Y} \supset \mathfrak{X}$, either $\mathfrak{Y}$ is $\mathfrak{X}$, or $A(\mathfrak{Y})=0$, or $A(\mathfrak{Y})>A(\mathfrak{X})$. This follows from the following formula for $A(\mathfrak{X})$ due to Cauchy:

CAUCHY'S SURFACE-AREA FORMULA:

$$A(\mathfrak{X}) = \int_\Omega V_{n-1}(\mathfrak{X};\, \mathbf{u})\, d\omega \Big/ V_{n-1}(\mathfrak{S}),$$

where $\mathfrak{S}$ *is an* $(n-1)$*-dimensional unit sphere,* $d\omega$ *is an element of surface of* Ω *the n-dimensional unit sphere,* $\mathbf{u}$ *is a variable unit vector and* $V_{n-1}(\mathfrak{X};\, \mathbf{u})$ *is the* $(n-1)$*-dimensional measure of the projection of* $\mathfrak{X}$ *in the direction* $\mathbf{u}$.

Since both sides of the formula are continuous functions of $\mathfrak{X}$ it is only necessary to consider the case when $\mathfrak{X}$ is a polytope. Let the $(n-1)$-dimensional faces of $\mathfrak{X}$ be $\mathfrak{X}_1, \mathfrak{X}_2, \ldots, \mathfrak{X}_N$, where $\mathfrak{X}_i$ is normal to the unit vector $\mathbf{v}_i$. For all directions $\mathbf{u}$ except those of a set of $(n-1)$-dimensional measure zero, the projection of the set $\mathfrak{X}$ is covered exactly twice by the projections of the sets $\mathfrak{X}_i$. Thus

$$\begin{aligned}\int_\Omega V_{n-1}(\mathfrak{X};\, \mathbf{u})\, d\omega &= \frac{1}{2}\sum_{i=1}^{N} \int_\Omega V_{n-1}(\mathfrak{X}_i;\, \mathbf{u})\, d\omega \\ &= \frac{1}{2}\sum_{i=1}^{N} V_{n-1}(\mathfrak{X}_i) \int_\Omega |\,\mathbf{u}\,.\,\mathbf{v}_i\,|\, d\omega.\end{aligned}$$

Now $$\int_\Omega |\,\mathbf{u}\,.\,\mathbf{v}_i\,|\, d\omega = 2V_{n-1}(\mathfrak{S});$$

thus the formula is correct.

EXERCISE 5·3

1. $F(\mathfrak{X})$ is a function that is defined for every closed bounded convex set $\mathfrak{X}$ and has the properties

(i) if $\mathfrak{X}$ is a polytope $F(\mathfrak{X})$ is equal to the sum of the $(n-1)$-dimensional measures of its faces;

(ii) if T is a transformation of R^n onto itself such that $|\,T(\mathbf{x}_1)-T(\mathbf{x}_2)\,| \leqslant |\,\mathbf{x}_1-\mathbf{x}_2\,|$, then $F(T(\mathfrak{X})) \leqslant F(\mathfrak{X})$.

Show that $F(\mathfrak{X})$ is identical with $A(\mathfrak{X})$.

4. Steiner symmetrization

We have already met the most important convexity preserving transformations, namely, the affine transformations. These transformations are applied to every point of the space concerned and transform each point into a second uniquely defined point. In contrast to this the transformations to be described in this section are not applied to individual points, they are applied to individual convex subsets of the space and transform each convex set into another convex set. A particular point may be transformed into different points according as it is considered to belong to different convex sets.

Let $\mathfrak{P}$ be a given hyperplane; we shall transform each convex set $\mathfrak{X}$ into another convex set $\mathfrak{X}_{\mathfrak{P}}$ by the following process. For each point $\mathbf{p}$ of $\mathfrak{P}$ let $\mathfrak{A}(\mathbf{p})$ be the line through $\mathbf{p}$ that is perpendicular to $\mathfrak{P}$. The set $\mathfrak{A}(\mathbf{p}) \cap \mathfrak{X}$ is either a closed segment or a point or the empty set. If it is a segment let $\mathfrak{B}(\mathbf{p})$ be the segment of equal length whose mid-point is at $\mathbf{p}$ and which lies in $\mathfrak{A}(\mathbf{p})$. If it is a point let $\mathfrak{B}(\mathbf{p})$ be the point $\mathbf{p}$. If it is the empty set let $\mathfrak{B}(\mathbf{p})$ be the empty set. Define $\mathfrak{X}_{\mathfrak{P}}$ by

$$\mathfrak{X}_{\mathfrak{P}} = \bigcup_{\mathbf{p} \in \mathfrak{P}} \mathfrak{B}(\mathbf{p}).$$

The process of obtaining $\mathfrak{X}_{\mathfrak{P}}$ from $\mathfrak{X}$ is known as *Steiner symmetrization* about the hyperplane $\mathfrak{P}$ and $\mathfrak{X}_{\mathfrak{P}}$ is called the *Steiner symmetral* of $\mathfrak{X}$ about $\mathfrak{P}$. $\mathfrak{X}$ is sometimes referred to as the *symmetrand.*

Steiner symmetrization has the following properties:

(i) *It transforms a convex set into a convex set.*

(ii) *It leaves the volume of the set unchanged.*

(iii) *It does not increase the surface area and will actually reduce it unless the original set is symmetric about a hyperplane parallel to the hyperplane of symmetrization.*†

(iv) *It does not increase the diameter of the set.*

(v) *It may reduce or it may increase the minimal width of a set.*

† Or lies completely in a hyperplane perpendicular to the hyperplane of symmetrization. This case is a trivial exception and we shall not consider it.

(vi) *If every member of a concave array is symmetrized about a given hyperplane the class of symmetrals again forms a concave array.*

(i), (ii) and (iv) are proved in the next theorem, Theorem 43. (vi) is proved in Theorem 44 and the first part of (iii) is a corollary of this theorem. Examples to illustrate (v) are given after Theorem 44. The second part of (iii), the case when the area is not actually reduced, is proved later, at the end of this paragraph, by an entirely different method.

THEOREM 43. *If $\mathfrak{X}$ is a closed bounded convex set, then $\mathfrak{X}_{\mathfrak{P}}$ is a closed bounded convex set and $V(\mathfrak{X}_{\mathfrak{P}}) = V(\mathfrak{X})$, $D(\mathfrak{X}_{\mathfrak{P}}) \leqslant D(\mathfrak{X})$.*

To show that $\mathfrak{X}_{\mathfrak{P}}$ is closed, let $\{\mathbf{x}_i\}$ be a sequence of points of $\mathfrak{X}_{\mathfrak{P}}$ convergent to the point $\mathbf{x}$. We shall show that $\mathbf{x} \in \mathfrak{X}_{\mathfrak{P}}$. Let the projection of $\mathbf{x}_i$ on $\mathfrak{P}$ be $\mathbf{p}_i$. The sequence $\{\mathbf{p}_i\}$ converges to a point of $\mathfrak{P}$, say $\mathbf{p}$, and $\overline{\lim\limits_{i\to\infty}}\, \lambda(\mathbf{p}_i) \leqslant \lambda(\mathbf{p})$,† where $\lambda(\mathbf{p}_i)$, $\lambda(\mathbf{p})$ are the lengths of $\mathfrak{B}(\mathbf{p}_i)$ and of $\mathfrak{B}(\mathbf{p})$. Now

$$|\mathbf{p}-\mathbf{x}| = \lim |\mathbf{p}_i - \mathbf{x}_i| \leqslant \tfrac{1}{2}\overline{\lim}\, \lambda(\mathbf{p}_i) \leqslant \tfrac{1}{2}\lambda(\mathbf{p});$$

thus either $\mathbf{x}$ is $\mathbf{p}$ or is on $\mathfrak{B}(\mathbf{p})$ and in either case $\mathbf{x} \in \mathfrak{X}_{\mathfrak{P}}$.

To see that $\mathfrak{X}_{\mathfrak{P}}$ is convex we remark that if $\mathfrak{Y} \subset \mathfrak{X}$ then $\mathfrak{Y}_{\mathfrak{P}} \subset \mathfrak{X}_{\mathfrak{P}}$. Let $\mathbf{x}_1, \mathbf{x}_2$ be two points of $\mathfrak{X}_{\mathfrak{P}}$ with projections $\mathbf{p}_1, \mathbf{p}_2$ on $\mathfrak{P}$. Let $\mathbf{x}_1'$ and $\mathbf{x}_2'$ be their reflexions in $\mathfrak{P}$ respectively. Let

$$\mathbf{x}_{\mathfrak{P}} = \mu_1\mathbf{x}_1 + \mu_2\mathbf{x}_2 \quad (\mu_1+\mu_2 = 1, \mu_1 \geqslant 0, \mu_2 \geqslant 0).$$

Then since $\mathfrak{X}$ is convex, for the point

$$\mathbf{p} = \mu_1\mathbf{p}_1 + \mu_2\mathbf{p}_2$$

we have
$$\lambda(\mathbf{p}) \geqslant \mu_1\lambda(\mathbf{p}_1) + \mu_2\lambda(\mathbf{p}_2).$$

Thus $\mathbf{x} \in \mathfrak{B}(\mathbf{p}) \subset \mathfrak{X}_{\mathfrak{P}}$ and $\mathfrak{X}_{\mathfrak{P}}$ is convex.

$V(\mathfrak{X}) = V(\mathfrak{X}_{\mathfrak{P}})$ is an immediate consequence of the definition of volume as meaning Lebesgue measure, for the projections of $\mathfrak{X}$ and $\mathfrak{X}_{\mathfrak{P}}$ are identical, and the line $\mathfrak{A}(\mathbf{p})$ meets $\mathfrak{X}$ and $\mathfrak{X}_{\mathfrak{P}}$ in two segments of equal length for each point $\mathbf{p}$.

† It is not necessarily true that $\lambda(\mathbf{p}_i) \to \lambda(\mathbf{p})$ as $i \to \infty$. For example, in R^3 consider a plane $\mathfrak{P}$ in which there is a circle $\mathfrak{C}$. Let $\mathfrak{l}$ be a segment which is perpendicular to $\mathfrak{P}$ and which passes through a point $\mathbf{p}$ of $\mathfrak{C}$. Denote the convex cover of $\mathfrak{l} \cup \mathfrak{C}$ by $\mathfrak{X}$. Let $\{\mathbf{p}_i\}$ be a sequence of points on $\mathfrak{C}$ converging to $\mathbf{p}$. Every line through $\mathbf{p}_i$ perpendicular to $\mathfrak{P}$ meets $\mathfrak{X}$ in a single point, so that $\lambda(\mathbf{p}_i) = 0$. But $\lambda(\mathbf{p})$ is equal to the length of $\mathfrak{l}$.

To show that $D(\mathfrak{X}_{\mathfrak{P}}) \leqslant D(\mathfrak{X})$ let $\mathbf{x}_1, \mathbf{x}_2$ be two points of $\mathfrak{X}_{\mathfrak{P}}$ for which $|\mathbf{x}_1 - \mathbf{x}_2| = D(\mathfrak{X}_{\mathfrak{P}})$. Let the lines through $\mathbf{x}_1$ and $\mathbf{x}_2$ respectively perpendicular to $\mathfrak{P}$ meet the frontier of $\mathfrak{X}$ in $\mathbf{x}_1', \mathbf{x}_1''$ and $\mathbf{x}_2', \mathbf{x}_2''$ respectively. Then $|\mathbf{x}_1 - \mathbf{x}_2|$ is not greater than the largest distance between two of the four points $\mathbf{x}_1', \mathbf{x}_1'', \mathbf{x}_2', \mathbf{x}_2''$.

This completes the proof of the theorem.

THEOREM 44. *If $\mathfrak{X}(\theta)$ is a concave array depending on a single parameter $\theta, 0 \leqslant \theta \leqslant 1$, then $[\mathfrak{X}(\theta)]_{\mathfrak{P}}$ is also a concave array.*

In R^{n+1} let $\mathfrak{Y}$ be the closed bounded convex set whose intersection with each hyperplane $x_{n+1} = \theta$ is $\mathfrak{X}(\theta)$ for $0 \leqslant \theta \leqslant 1$. Let $\mathfrak{Q}$ be the n-dimensional space spanned by $\mathfrak{P}$ and the x_{n+1} axis. Symmetrize $\mathfrak{Y}$ about $\mathfrak{Q}$, then the intersection of $\mathfrak{Y}_{\mathfrak{Q}}$ with $x_{n+1} = \theta$ projects into $[\mathfrak{X}(\theta)]_{\mathfrak{P}}$ on $x_{n+1} = 0$. Thus $[\mathfrak{X}(\theta)]_{\mathfrak{P}}$ is a concave array.

COROLLARY. $A(\mathfrak{X}_{\mathfrak{P}}) \leqslant A(\mathfrak{X})$.

Consider the array $(1-\theta)\mathfrak{X} + \theta\mathfrak{S}$, where $\mathfrak{S}$ is a unit sphere whose centre lies on $\mathfrak{P}$. Then $\mathfrak{S}_{\mathfrak{P}} = \mathfrak{S}$. As the array

$$[(1-\theta)\mathfrak{X} + \theta\mathfrak{S}]_{\mathfrak{P}}$$

is concave, we have

$$[(1-\theta)\mathfrak{X} + \theta\mathfrak{S}]_{\mathfrak{P}} \supset (1-\theta)\mathfrak{X}_{\mathfrak{P}} + \theta\mathfrak{S}.$$

Thus
$$\left[\mathfrak{X} + \frac{\theta}{1-\theta}\mathfrak{S}\right]_{\mathfrak{P}} \supset \mathfrak{X}_{\mathfrak{P}} + \frac{\theta}{1-\theta}\mathfrak{S},$$

and on writing δ for $\theta/(1-\theta)$ and using $V(\mathfrak{X}_{\mathfrak{P}}) = V(\mathfrak{X})$ we have

$$V(\mathfrak{U}(\mathfrak{X}, \delta)) - V(\mathfrak{X}) \geqslant V(\mathfrak{U}(\mathfrak{X}_{\mathfrak{P}}, \delta)) - V(\mathfrak{X}_{\mathfrak{P}}).$$

Hence
$$A(\mathfrak{X}) \geqslant A(\mathfrak{X}_{\mathfrak{P}}).$$

Two examples in R^2:

(*a*) $\mathfrak{X}_1$ is a non-rectangular parallelogram, whose shorter side is smaller than its width in a perpendicular direction.

(*b*) $\mathfrak{X}_2$ is a Reuleaux triangle, i.e. the set bounded by three circular arcs AB, BC, CA, each of radius say λ, where ABC is an equilateral triangle of side length λ.

Symmetrize $\mathfrak{X}_1$ about a line perpendicular to its shorter side. We obtain a rectangle whose minimal width is its shorter side. This is greater than the minimal width of $\mathfrak{X}_1$.

Symmetrize $\mathfrak{X}_2$ about AB to obtain $[\mathfrak{X}_2]_{\mathfrak{P}}$. Since $\mathfrak{X}_2$ is not symmetric about any line parallel to AB, the effect is to reduce the length of the perimeter of the set $\mathfrak{X}_2$. Now the length of perimeter of $\mathfrak{X}_2$ is (by Cauchy's formula § 5·3)

$$\frac{1}{2}\int_0^{2\pi} w(\mathfrak{X}_2, \theta)\, d\theta,$$

where $w(\mathfrak{X}_2, \theta)$ is the width of $\mathfrak{X}_2$ in the direction θ. Thus we have

$$\int_0^{2\pi} w([\mathfrak{X}_2]_{\mathfrak{P}}, \theta)\, d\theta < \int_0^{2\pi} w(\mathfrak{X}_2, \theta)\, d\theta = 2\pi\lambda,$$

and it follows that the minimal width of $[\mathfrak{X}_2]_{\mathfrak{P}}$ is less than λ which is itself the minimal width of $\mathfrak{X}_2$.

Steiner symmetrization when applied to a linear array will always produce a concave array which may or may not be linear. In the next theorem we consider the conditions under which the symmetrized array will be linear.

THEOREM 45. *The linear array* $(1-\theta)\mathfrak{X}_1 + \theta\mathfrak{X}_2$, $0 \leqslant \theta \leqslant 1$, *symmetrizes into a linear array about every hyperplane* $\mathfrak{P}$ *if and only if* $\mathfrak{X}_1$ *and* $\mathfrak{X}_2$ *are similar and similarly situated convex sets.*

It is obvious that the array symmetrizes into a linear array if $\mathfrak{X}_1$ and $\mathfrak{X}_2$ are similar and the proof of this is omitted.

Let $\mathfrak{Y}$ be the closed bounded convex subset of R^{n+1} such that the hyperplane $x_{n+1} = \theta$ intersects $\mathfrak{Y}$ in a set whose projection on $x_{n+1} = 0$ is $(1-\theta)\mathfrak{X}_1 + \theta\mathfrak{X}_2$. Let $\mathfrak{Q}$ be the n-dimensional space spanned by $\mathfrak{P}$ in $x_{n+1} = 0$ and the x_{n+1} axis. If $[(1-\theta)\mathfrak{X}_1 + \theta\mathfrak{X}_2]_{\mathfrak{P}}$ is linear, then $\mathfrak{Y}_{\mathfrak{Q}}$ must be the convex cover of its intersections with $x_{n+1} = 0$ and with $x_{n+1} = 1$. Denote these by $\mathfrak{Y}_{0,\mathfrak{Q}}$ and $\mathfrak{Y}_{1,\mathfrak{Q}}$, respectively, and let $\mathfrak{Y}_0$ and $\mathfrak{Y}_1$ be the intersections of $\mathfrak{Y}$ with $x_{n+1} = 0$ and with $x_{n+1} = 1$.

Select a point $\mathbf{p}_{0,\mathfrak{Q}}$ of the relative frontier of $\mathfrak{Y}_{0,\mathfrak{Q}}$ and let $\mathbf{p}_{1,\mathfrak{Q}}$ be a point on the relative frontier of $\mathfrak{Y}_{1,\mathfrak{Q}}$ which lies on a support hyperplane to $\mathfrak{Y}_{\mathfrak{Q}}$ at $\mathbf{p}_{0,\mathfrak{Q}}$. The two-dimensional space $\mathfrak{R}$ which includes $\mathbf{p}_{0,\mathfrak{Q}}$ and $\mathbf{p}_{1,\mathfrak{Q}}$ and is perpendicular to $\mathfrak{Q}$ meets $\mathfrak{Y}_{\mathfrak{Q}}$ in a trapezium $\mathfrak{T}_{\mathfrak{Q}}$. Further, it meets $\mathfrak{Y}$ in a trapezium $\mathfrak{T}$ of which $\mathfrak{T}_{\mathfrak{Q}}$ is the symmetral. Let the vertices of $\mathfrak{T}$ be, in order, $\mathbf{p}_0\mathbf{p}_1\mathbf{p}_1'\mathbf{p}_0'$, where $\mathbf{p}_0, \mathbf{p}_0'$ belong to $\mathfrak{Y}_0$ and $\mathbf{p}_1, \mathbf{p}_1'$ belong to $\mathfrak{Y}_1$. Then $\mathbf{p}_0, \mathbf{p}_1$ lie on the same support hyperplane to $\mathfrak{Y}$ and so do $\mathbf{p}_0', \mathbf{p}_1'$.

Thus if the support hyperplanes to $\mathfrak{X}_1$ and $\mathfrak{X}_2$ at $\mathbf{x}_1$ and $\mathbf{x}_2$ on $\operatorname{Fr}\mathfrak{X}_1$ and $\operatorname{Fr}\mathfrak{X}_2$ respectively are parallel and similarly situated, and if the line through $\mathbf{x}_i$ perpendicular to $\mathfrak{P}$ meets $\mathfrak{X}_i$ again in $\mathbf{y}_i$, then the support hyperplane to $\mathfrak{X}_1$ at $\mathbf{y}_1$ is parallel to that to $\mathfrak{X}_2$ at $\mathbf{y}_2$.

Since we may select $\mathbf{x}_1, \mathbf{x}_2$ to be exposed points of $\mathfrak{X}_1$ and $\mathfrak{X}_2$ respectively (see Exercise 5·1, 4), and since we may give to $\mathfrak{P}$ any orientation, it follows, that $\mathfrak{X}_1$ and $\mathfrak{X}_2$ are similar and similarly situated.

All the properties that we stated concerning Steiner symmetrization have now been proved except for the case of equality between the surface areas of the symmetral and the symmetrand. To establish that equality only holds when the symmetrand is a translation of the symmetral we need to consider a different range of ideas. This result is more difficult to prove than the others and we give here an outline only.

Let $\mathfrak{X}$ be a convex body whose projection on the hyperplane $x_n = 0$ is the set $\mathfrak{X}_0$. We shall consider first that case in which $\mathfrak{X}_0$ is a polytope. We can define $\mathfrak{X}$ by means of two functions $f(x_1, \ldots, x_{n-1})$ and $g(x_1, \ldots, x_{n-1})$ defined over $\mathfrak{X}_0$ as

$$\mathfrak{X} = \{(x_1, x_2, \ldots, x_n); (x_1, x_2, \ldots, x_{n-1}, 0) \in \mathfrak{X}_0, \quad g(x_1, \ldots, x_{n-1}) \leqslant x_n \leqslant f(x_1, \ldots, x_{n-1})\}. \tag{5·4·1}$$

The functions f and g are concave and convex respectively. Let $\mathbf{e}_i$ be the vector all of whose elements are zero save the ith which is unity and write

$$f_i(x_1, \ldots, x_{n-1}) = \delta f((x_1, \ldots, x_{n-1}); \mathbf{e}_i),$$
$$g_i(x_1, \ldots, x_{n-1}) = \delta g((x_1, \ldots, x_{n-1}); \mathbf{e}_i),$$

where $1 \leqslant i \leqslant n-1$. Then

$$A(\mathfrak{X}) = \int_{\mathrm{Fr}(\mathfrak{X}_0)} (f-g)\, d\sigma_1 + \int_{\mathfrak{X}_0} \left[\left\{1 + \sum_{i=1}^{n-1} f_i^2\right\}^{\frac{1}{2}} + \left\{1 + \sum_{i=1}^{n-1} g_i^2\right\}^{\frac{1}{2}}\right] d\sigma_2, \tag{5·4·2}$$

where $d\sigma_1$ and $d\sigma_2$ are elements of $(n-2)$ and $(n-1)$-dimensional volume and we have omitted the arguments in f, g, etc.

The formula (5·4·2) is easily proved when $\mathfrak{X}$ is a polytope. When $\mathfrak{X}$ is not a polytope, then (since $\mathfrak{X}_0$ is a polytope) we can define a sequence of polytopes $\mathfrak{X}^{(j)} \to \mathfrak{X}$ such that if $f^{(j)}$ and $g^{(j)}$ are functions defined over $\mathfrak{X}_0$ as are f, g in (5·4·1), but using $\mathfrak{X}^{(j)}$ instead of $\mathfrak{X}$, we have the following properties:

(i)
$$f^{(j)}(x_1, \ldots, x_{n-1}) \to f(x_1, \ldots, x_{n-1}),$$
$$g^{(j)}(x_1, \ldots, x_{n-1}) \to g(x_1, \ldots, x_{n-1}),$$

where $(x_1, \ldots, x_{n-1}, 0)$ belongs to $\mathfrak{X}_0$.

(ii) The projection of $\mathfrak{X}^{(j)}$ on $x_n = 0$ is $\mathfrak{X}_0$, $j = 1, 2, \ldots$.

We shall show that under these circumstances

$$f_i^{(j)} \to f_i \quad \text{and} \quad g_i^{(j)} \to g_i \quad (1 \leqslant i \leqslant n-1)$$

as $j \to \infty$ for all points $(x_1, \ldots, x_{n-1}, 0)$ except at most a set of measure zero in dimension $n-1$. Consider a segment

$$(x_1, \ldots, x_{n-1}, 0), \quad (x_1', x_2, \ldots, x_{n-1}, 0) \quad (x_1' > x_1),$$

contained in $\mathfrak{X}_0$ and parallel to the x_1 axis. For each j,

$$f^{(j)}(x_1 + t, x_2, \ldots, x_{n-1}, 0)$$

is an absolutely continuous function of t, $0 \leqslant t \leqslant x_1' - x_1$ and $f_1^{(j)}$ is a decreasing function of t. Thus for every c, $0 \leqslant c \leqslant x_1' - x_1$, we have

$$f^{(j)}(x_1 + c, x_2, \ldots, x_{n-1}) - f^{(j)}(x_1, x_2, \ldots, x_{n-1})$$
$$= \int_0^c f_1^{(j)}(x_1 + t, x_2, \ldots, x_{n-1})\, dt. \qquad (5·4·3)$$

Now the left-hand side of this inequality tends to

$$f(x_1 + c, x_2, \ldots, x_{n-1}) - f(x_1, x_2, \ldots, x_{n-1})$$

as $j \to \infty$ for all c, $0 \leqslant c \leqslant x_1' - x_1$. If the segment belongs to the relative interior of $\mathfrak{X}_0$, the functions $f_1^{(j)}(x_1 + t, x_2, \ldots, x_{n-1})$ are uniformly bounded. Finally, these functions are all decreasing. Therefore

$$f_1^{(j)}(x_1 + t, x_2, \ldots, x_{n-1}) \to f_1(x_1 + t, x_2, \ldots, x_{n-1})$$

for almost all t, $0 \leqslant t \leqslant x_1' - x_1$. Taking any segment parallel to the x_1 axis joining two points of the relative interior of $\mathfrak{X}_0$ we see that $f_1^{(j)} \to f_1$ except for points belonging to a set of $(n-1)$-dimensional measure zero. A similar argument holds for the limiting relation $f_i^{(j)} \to f_i$, $1 \leqslant i \leqslant n-1$.

Using (5·4·2) with $\mathfrak{X}^{(j)}, f^{(j)}, g^{(j)}$ in the place of $\mathfrak{X}, f, g$ and letting $j \to \infty$, we obtain the required result.

Now symmetrize $\mathfrak{X}$ about $x_n = 0$ to produce $\mathfrak{X}_{\mathfrak{P}}$. The functions f and g are replaced by $\frac{f-g}{2}$ and $\frac{g-f}{2}$. Thus from (5·4·2) we have

$$A(\mathfrak{X}_{\mathfrak{P}}) = \int_{\mathrm{Fr}\,\mathfrak{X}_0} (f-g)\,d\sigma_1 + \int_{\mathfrak{X}_0} \left[\left\{1+\sum_1^{n-1}\left(\frac{f_i-g_i}{2}\right)^2\right\}^{\frac{1}{2}} + \left\{1+\sum_1^{n-1}\left(\frac{g_i-f_i}{2}\right)^2\right\}^{\frac{1}{2}}\right] d\sigma_2. \quad (5\cdot4\cdot4)$$

By a well-known inequality $A(\mathfrak{X}_{\mathfrak{P}}) < A(\mathfrak{X})$ unless $f_i = -g_i$ for almost all points of $\mathfrak{X}_0$ (in the sense of $(n-1)$-dimensional measure). But then relations involving f and f_i similar to (5·4·3) show that for almost all $(x_1 \dots x_{n-1}, 0)$ of $\mathfrak{X}_0$ $f = -g+k$, where k is a constant. Now f and g are continuous. Thus $f = -g+k$ for all $(x_1, \dots, x_{n-1})$ of $\mathfrak{X}_0$. Thus $\mathfrak{X}$ is symmetric about a hyperplane parallel to $x_n = 0$.

So far we have considered only the case when $\mathfrak{X}$ is a convex body whose projection on $x_n = 0$ is a polytope. If $\mathfrak{X}$ is any convex body which is not symmetric about a hyperplane parallel to the hyperplane of symmetrization $x_n = 0$, we can select a finite number of points in the projection of $\mathfrak{X}$ on $x_n = 0$ forming the vertices of a polytope $\mathfrak{Y}_0$, and then construct the part of $\mathfrak{X}$ that projects onto $\mathfrak{Y}_0$. We take $\mathfrak{Y}_0$ to have interior points relative to $x_n = 0$. By considering (5·4·2) for $\mathfrak{Y}$ and using the fact that the area of the set $\mathfrak{X} \dot{-} \mathfrak{Y}$ (defined by the Minkowski procedure, see equation (5·3·1)) is not increased by symmetrization, we see that $A(\mathfrak{X})$ is in fact strictly decreased by symmetrization.

Thus property (iii) of Steiner symmetrization is completely proved.

5. The Brünn-Minkowski theorem: the Minkowski and Fenchel-Alexandroff inequalities

If we consider a plane convex set whose projection on the x axis is the closed segment $0 \leqslant x \leqslant 1$, then any line through x perpendicular to the segment meets the convex set in a segment of length say $\lambda(x)$ and $\lambda(x)$ is a concave function of x. We may interpret the result in a different manner, namely, that if we are given in a line a concave array of segments depending on a single parameter, then the length of the segments is a concave function of the parameter. This is the one-dimensional form of the Brünn-Minkowski inequality.

THEOREM 46. *If* $\mathfrak{X}_\theta = (1-\theta)\mathfrak{X}_0 + \theta\mathfrak{X}_1$ *then the n-th root of* $V(\mathfrak{X}_\theta)$ *is a concave function of* θ.

We give two proofs of this result. The first applies to non-convex sets, whilst the second, although not immediately applicable to non-convex sets, gives the conditions for equality for convex sets.

First proof. Suppose that each of $\mathfrak{X}_0$ and $\mathfrak{X}_1$ consists of a finite number of parallelepipeds whose faces are parallel to the co-ordinate hyperplanes. Suppose further that no two parallelepipeds of $\mathfrak{X}_0$ intersect except in an at most $(n-1)$-dimensional set, and similarly that no two parallelepipeds of $\mathfrak{X}_1$ intersect except in an at most $(n-1)$-dimensional set. The proof is based on an induction with respect to the total number of parallelepipeds in $\mathfrak{X}_0$ and $\mathfrak{X}_1$ combined.

If $\mathfrak{X}_0$ and $\mathfrak{X}_1$ are each single parallelopipeds the inequality can be proved by direct calculation. We assume that the result has been proved if the total number of parallelepipeds in $\mathfrak{X}_1$ and $\mathfrak{X}_0$ combined is less than m. Now consider the case when there are exactly m parallelepipeds in $\mathfrak{X}_1$ and $\mathfrak{X}_0$. We may suppose that one of $\mathfrak{X}_0$, $\mathfrak{X}_1$ (say $\mathfrak{X}_0$) contains at least two parallelepipeds. There is a hyperplane parallel to one of the coordinate hyperplanes, say $x_i = \lambda$, such that at least one parallelepiped of $\mathfrak{X}_0$ has no points in $x_i > \lambda$ and at least one has no points in $x_i < \lambda$. Denote the set of parallelepipeds and parts of parallelepipeds of $\mathfrak{X}_0$ that lie in $x_i < \lambda$ by $\mathfrak{X}_0^-$, that in $x_i > \lambda$ by $\mathfrak{X}_0^+$. There is a hyperplane $x_i = \mu$ that divides $\mathfrak{X}_1$ similarly into two sets $\mathfrak{X}_1^-$ $\mathfrak{X}_1^+$ such that

$$\frac{V(\mathfrak{X}_0^+)}{V(\mathfrak{X}_0^-)} = \frac{V(\mathfrak{X}_1^+)}{V(\mathfrak{X}_1^-)}. \tag{5·5·1}$$

Applying the induction hypothesis to $\mathfrak{X}_0^+$, $\mathfrak{X}_1^+$ and to $\mathfrak{X}_0^-$, $\mathfrak{X}_1^-$, and observing that $(1-\theta)\mathfrak{X}_0^- + \theta\mathfrak{X}_1^-$ is disjoint from $(1-\theta)\mathfrak{X}_0^+ + \theta\mathfrak{X}_1^+$ we obtain

$$\begin{aligned} V(\mathfrak{X}_\theta) &\geqslant V((1-\theta)\mathfrak{X}_0^- + \theta\mathfrak{X}_1^-) + V((1-\theta)\mathfrak{X}_0^+ + \theta\mathfrak{X}_1^+) \\ &\geqslant [(1-\theta)\,V(\mathfrak{X}_0^-)^{1/n} + \theta V(\mathfrak{X}_1^-)^{1/n}]^n \\ &\quad + [(1-\theta)\,V(\mathfrak{X}_0^+)^{1/n} + \theta V(\mathfrak{X}_1^+)^{1/n}]^n \\ &= [(1-\theta)\,V(\mathfrak{X}_0)^{1/n} + \theta V(\mathfrak{X}_1)^{1/n}]^n \end{aligned}$$

by (5·5·1).

Thus the next stage of the induction is complete.

It follows by a standard approximation argument that the Brünn-Minkowski theorem holds for any measurable sets.

Second proof. We first select n hyperplanes $\mathfrak{P}_1, \ldots, \mathfrak{P}_n$ that are perpendicular in pairs and intersect in a point say $\mathbf{p}$. Symmetrize the concave array $\mathfrak{X}_\theta$ successively about $\mathfrak{P}_1, \ldots, \mathfrak{P}_n$. We obtain a concave array which we may denote by $\mathfrak{Y}_\theta^{(0)}$ with the property that each member of $\mathfrak{Y}_\theta^{(0)}$ is central with the point $\mathbf{p}$ as its centre.

Select n linearly independent hyperplanes $\mathfrak{Q}_0, \ldots, \mathfrak{Q}_{n-1}$ through $\mathbf{p}$ which are such that the perpendiculars to them in pairs make angles which are irrational multiples of π. Make an infinite sequence of successive symmetrizations of $\mathfrak{Y}_\theta^{(0)}$, where the ith symmetrization is about $\mathfrak{Q}_j$, where j is the residue of $i \bmod n$. Denote the resulting sequence of arrays by

$$\mathfrak{Y}_\theta^{(0)}, \mathfrak{Y}_\theta^{(1)}, \mathfrak{Y}_\theta^{(2)}, \ldots.$$

All the sets of all these arrays are uniformly bounded. For the three particular values of θ, 0, 1 and δ we select a subsequence of integers i_k such that $\mathfrak{Y}_0^{(i_k)}, \mathfrak{Y}_1^{(i_k)}, \mathfrak{Y}_\delta^{(i_k)}$ converge respectively to, say, $\mathfrak{Y}_0, \mathfrak{Y}_1, \mathfrak{Y}_\delta$. Each of these last three sets is, in fact, a sphere. Consider, for example, $\mathfrak{Y}_0$. There is at least one integer r such that $i_k \equiv r \bmod n$ has infinitely many solutions in k and $\mathfrak{Y}_0$ is symmetric about $\mathfrak{Q}_r$. Now consider the sequence $\mathfrak{Y}_0^{(i_k+1)}$; this converges to a set say $\mathfrak{Z}$ and $\mathfrak{Z}$ is symmetric about $\mathfrak{Q}_{r+1}$ or about $\mathfrak{Q}_0$ if $r = n-1$. $\mathfrak{Z}$ is the symmetral of $\mathfrak{Y}_0$ about $\mathfrak{Q}_{r+1}$ (or $\mathfrak{Q}_0$ if $r = n-1$). Thus if $\mathfrak{Y}_0$ is not symmetric about a hyperplane parallel to $\mathfrak{Q}_{r+1}$ then

$$A(\mathfrak{Z}) < A(\mathfrak{Y}_0).$$

But

$$A(\mathfrak{Y}_0^{i_{k+1}}) \leqslant A(\mathfrak{Y}_0^{i_k+1});$$

thus in the limit as $k \to \infty$

$$A(\mathfrak{Y}_0) \leqslant A(\mathfrak{Z}). \tag{5·5·2}$$

Since this is a contradiction it follows that $\mathfrak{Y}_0$ is symmetric about a hyperplane parallel to $\mathfrak{Q}_{r+1}$. But $\mathfrak{Y}_0$ has $\mathbf{p}$ as its centre and the segment belonging to $\mathfrak{Y}_0$ of the line through $\mathbf{p}$ perpendicular to $\mathfrak{Q}_{r+1}$ has its mid-point at $\mathbf{p}$, i.e. at a point on $\mathfrak{Q}_{r+1}$. Thus $\mathfrak{Y}_0$ is symmetric about $\mathfrak{Q}_{r+1}$. Similarly, it is symmetric about $\mathfrak{Q}_{r+2}$ and in fact about each $\mathfrak{Q}_i$, $i = 0, 1, \ldots, n-1$.

But then as these hyperplanes make irrational angles with one another it follows that $\mathfrak{Y}_0$ is a sphere with $\mathbf{p}$ as its centre. Moreover, at each symmetrization the array $\mathfrak{Y}_\theta^{(i)}$ is concave. Thus $\mathfrak{Y}_\delta$ is a sphere which contains the concentric sphere $(1-\delta)\mathfrak{Y}_0+\delta\mathfrak{Y}_1$. Denote the radius of $\mathfrak{Y}_\theta$ by r_θ for $\theta=0, 1, \delta$, then

$$r_\delta \geqslant (1-\delta)\, r_0+\delta r_1,$$

i.e. $$[V(\mathfrak{Y}_\delta)]^{1/n} \geqslant (1-\delta)\,[V(\mathfrak{Y}_0)]^{1/n}+\delta[V(\mathfrak{Y}_1)]^{1/n},$$

i.e. since $$V(\mathfrak{X}_\delta)=V(\mathfrak{Y}_\delta), \quad \text{etc.},$$

$$[V(\mathfrak{X}_\delta)]^{1/n} \geqslant (1-\delta)\,[V(\mathfrak{X}_0)]^{1/n}+\delta[V(\mathfrak{X}_1)]^{1/n},$$

and the Brünn-Minkowski inequality is proved.

Remark. Equality holds if and only if the two sets $\mathfrak{X}_0$ and $\mathfrak{X}_1$ are either homothetic or if they lie in parallel hyperplanes.

For if $V(\mathfrak{X}_0)=V(\mathfrak{X}_1)=0$ and $\mathfrak{X}_0$, $\mathfrak{X}_1$ do not lie in parallel hyperplanes then $\mathfrak{X}_\theta$ has interior points, $V(\mathfrak{X}_\theta)>0$, and strict inequality holds in the Brünn-Minkowski inequality.

Suppose $V(\mathfrak{X}_0)=0$ and $V(\mathfrak{X}_1)>0$. We shall consider a single point to be homothetic with any convex set, so that if $\mathfrak{X}_0$ is a single point, $\mathfrak{X}_0$ and $\mathfrak{X}_1$ are homothetic. Otherwise select two distinct points $\mathbf{x}', \mathbf{x}''$ of $\mathfrak{X}_0$. Let $\mathfrak{X}'_\theta$, $\mathfrak{X}''_\theta$ be the two sets $(1-\theta)\mathbf{x}'+\theta\mathfrak{X}_1$, $(1-\theta)\,\mathbf{x}''+\theta\mathfrak{X}_1$. Then

$$\mathfrak{X}_\theta \supset \mathfrak{X}'_\theta \cup \mathfrak{X}''_\theta$$

and $$V(\mathfrak{X}'_\theta)=V(\mathfrak{X}''_\theta)=\theta^n V(\mathfrak{X}_1).$$

Since $\mathfrak{X}'_\theta$ does not coincide with $\mathfrak{X}''_\theta$

$$V(\mathfrak{X}_\theta)>\theta^n V(\mathfrak{X}_1),$$

and strict inequality holds in the Brünn-Minkowski inequality

Similarly, if $V(\mathfrak{X}_1)=0$ and $V(\mathfrak{X}_0)>0$ we have strict inequality.

Finally, suppose that $V(\mathfrak{X}_1)>0$ and $V(\mathfrak{X}_0)>0$. The sets $(1-\theta)\,\mathfrak{X}_0+\theta\mathfrak{X}_1$ symmetrize into a linear array about every hyperplane only if $\mathfrak{X}_0$ and $\mathfrak{X}_1$ are homothetic, by Theorem 45. If $\mathfrak{X}_0$ and $\mathfrak{X}_1$ are not homothetic, then in the process described in the proof of Theorem 46 we can select the hyperplane $\mathfrak{Q}_0$ such that $\mathfrak{Y}_\theta^{(0)}$ is not linear. Then

$$V(\mathfrak{Y}_\delta^{(0)})=V((1-\delta)\,\mathfrak{Y}_0^{(0)}+\delta\mathfrak{Y}_1^{(0)})+\epsilon \quad (\epsilon>0). \qquad (5{\cdot}5{\cdot}3)$$

Hence by applying the result of Theorem 46 to the right-hand side of (5·5·3) we have the strict inequality

$$[V(\mathfrak{X}_\delta)]^{1/n} > (1-\delta)[V(\mathfrak{X}_0)]^{1/n} + \delta[V(\mathfrak{X}_1)]^{1/n},$$

and this completes the discussion of the cases of inequality.

The Brünn-Minkowski theorem can be stated as follows. Let $\mathfrak{P}_1, \mathfrak{P}_2$ be two parallel support hyperplanes of a convex set $\mathfrak{X}$ and let $\mathfrak{l}$ be a line perpendicular to them. In any hyperplane $\mathfrak{P}$ between $\mathfrak{P}_1$ and $\mathfrak{P}_2$ construct the $(n-1)$-dimensional solid sphere whose volume is $V_{n-1}(\mathfrak{P} \cap \mathfrak{X})$ and whose centre lies on $\mathfrak{l}$. The union of all these sets is denoted by $\mathfrak{Y}$. Then the Brünn-Minkowski theorem implies that $\mathfrak{Y}$ is a convex set. Of course other sets could be used in place of spheres. The process when spheres are used is sometimes called the Schwarz rotation process. $\mathfrak{Y}$ in this case is a solid of rotation with $\mathfrak{l}$ as its axis of rotation.

The Brünn-Minkowski theorem can be used to obtain inequalities for the mixed volumes of a linear array defined over two convex bodies, $\mathfrak{X}_\theta = (1-\theta)\mathfrak{X}_0 + \theta\mathfrak{X}_1$. We have

$$V(\mathfrak{X}_\theta) = \sum_{\nu=0}^{n} \binom{n}{\nu} (1-\theta)^{n-\nu}\theta^\nu V(\mathfrak{X}_0, n-\nu; \mathfrak{X}_1, \nu),$$

and since $[V(\mathfrak{X}_\theta)]^{1/n}$ is a concave function of θ that is linear only when $\mathfrak{X}_0$ and $\mathfrak{X}_1$ are homothetic it follows that the same is true of the function

$$f(\theta) = [V(\mathfrak{X}_\theta)]^{1/n} - (1-\theta)[V(\mathfrak{X}_0, n; \mathfrak{X}_1, 0)]^{1/n} - \theta[V(\mathfrak{X}_0, 0; \mathfrak{X}_1, n)]^{1/n}.$$

Since $f(0) = f(1) = 0$ and $f(\theta)$ is concave, it follows that $(df/d\theta)_{\theta=0}$ is non-negative and is zero only if $f(\theta) \equiv 0$. Thus we have

$$[V(\mathfrak{X}_0, n-1; \mathfrak{X}_1, 1)]^n \geq [V(\mathfrak{X}_0, n; \mathfrak{X}_1, 0)]^{n-1}[V(\mathfrak{X}_0, 0; \mathfrak{X}_1, n)], \quad (5·5·4)$$

and equality holds if and only if the sets $\mathfrak{X}_0$ and $\mathfrak{X}_1$ are homothetic. Similarly by interchanging $\mathfrak{X}_0$ and $\mathfrak{X}_1$ we obtain

$$[V(\mathfrak{X}_0, 1; \mathfrak{X}_1, n-1)]^n \geq [V(\mathfrak{X}_0, n; \mathfrak{X}_1, 0)][V(\mathfrak{X}_0, 0; \mathfrak{X}_1, n)]^{n-1}. \quad (5·5·5)$$

A second inequality may be found by considering the second derivative of $[V(\mathfrak{X}_\theta)]^{1/n}$ at $\theta = 0$. We obtain

$$[V(\mathfrak{X}_0, 1; \mathfrak{X}_1, n-1)]^2 \geq [V(\mathfrak{X}_0, n; \mathfrak{X}_1, 0)][V(\mathfrak{X}_0, n-2; \mathfrak{X}_1, 2)], \quad (5·5·6)$$

and, interchanging $\mathfrak{X}_0$ and $\mathfrak{X}_1$,

$$[V(\mathfrak{X}_0, n-1; \mathfrak{X}_1, 1)]^2 \geqslant [V(\mathfrak{X}_0, 2; \mathfrak{X}_1, n-2)][V(\mathfrak{X}_0, 0; \mathfrak{X}_1, n)]. \tag{5·5·7}$$

The inequalities (5·5·4)–(5·5·7) are known as Minkowski's inequalities; (5·5·6) and (5·5·7) are particular cases of

$$[V(\mathfrak{X}_0, \nu; \mathfrak{X}_1, n-\nu)]^2 \geqslant [V(\mathfrak{X}_0, \nu-1; \mathfrak{X}_1, n-\nu+1)] \times [V(\mathfrak{X}_0, \nu+1; \mathfrak{X}_1, n-\nu-1)], \tag{5·5·8}$$

which is contained in the Fenchel-Alexandroff inequality

$$[V(\mathfrak{X}_1, \mathfrak{X}_2, \mathfrak{X}_3, \ldots, \mathfrak{X}_n)]^2 \geqslant [V(\mathfrak{X}_1, \mathfrak{X}_1, \mathfrak{X}_3, \ldots, \mathfrak{X}_n)] \times [V(\mathfrak{X}_2, \mathfrak{X}_2, \mathfrak{X}_3, \ldots, \mathfrak{X}_n)], \tag{5·5·9}$$

where $\mathfrak{X}_1, \mathfrak{X}_2, \mathfrak{X}_3, \ldots, \mathfrak{X}_n$ are n convex bodies in R^n.

For the proof the reader is referred to the papers by Fenchel and Alexandroff.

6. Central symmetrization

Let $H(\mathbf{u})$ be the support function of a convex set $\mathfrak{X}$. Then the function $\frac{1}{2}[H(\mathbf{u})+H(-\mathbf{u})]$ is again the support function of some convex set say $\mathfrak{X}'$. We say that $\mathfrak{X}'$ is obtained from $\mathfrak{X}$ by central symmetrization. The width of $\mathfrak{X}$ in any direction is equal to that of $\mathfrak{X}'$ in the same direction. In particular, $\mathfrak{X}$ and $\mathfrak{X}'$ have the same diameter and minimal width. Also if we denote by $\mathfrak{X}_1$ the reflexion of $\mathfrak{X}$ in the origin then $\mathfrak{X}' = \frac{1}{2}(\mathfrak{X}+\mathfrak{X}_1)$ and by the Brünn-Minkowski inequality $V(\mathfrak{X}') \geqslant V(\mathfrak{X})$.

The set $2\mathfrak{X}'$ is also called the vector domain or difference set of $\mathfrak{X}$ because if each vector whose two end-points belong to $\mathfrak{X}$ is translated until its initial end-point lies at the origin then the set of final end-points is precisely the set $2\mathfrak{X}'$. We have

$$V(2\mathfrak{X}') \geqslant 2^n V(\mathfrak{X}). \tag{5·6·1}$$

The best possible inequality in the opposite sense is

$$V(2\mathfrak{X}') \leqslant \binom{2n}{n} V(\mathfrak{X})$$

and this has recently been established by Professor C. A. Rogers and Dr G. C. Shephard as follows:

Denote a point of R^{2n} by $(\mathbf{x}, \mathbf{y})$ where

$$\mathbf{x} = (x_1, \ldots, x_n), \quad \mathbf{y} = (y_1, \ldots, y_n)$$

and where the ith component of $(\mathbf{x}, \mathbf{y})$ is x_i if $1 \leqslant i \leqslant n$ and is y_{i-n} if $n < i \leqslant 2n$. Let $\mathfrak{Y}_1$, $\mathfrak{Y}_2$ be two n-dimensional convex sets each containing $\mathbf{O}$ and write

$$K(\mathbf{x}') = \{(\mathbf{x}, \mathbf{y});\ \mathbf{x} = \mathbf{x}'\}, \quad J(\mathbf{y}') = \{(\mathbf{x}, \mathbf{y});\ \mathbf{y} = \mathbf{y}'\},$$
$$\mathfrak{Z} = \{(\mathbf{x}, \mathbf{y});\ \mathbf{x} \in \mathfrak{Y}_1 \text{ and } \mathbf{x} + \mathbf{y} \in \mathfrak{Y}_2\}.$$

Then if $\mathbf{x}' \notin \mathfrak{Y}_1$ $\qquad K(\mathbf{x}') \cap \mathfrak{Z} = \phi,$

if $\mathbf{x}' \in \mathfrak{Y}_1$ $\qquad K(\mathbf{x}') \cap \mathfrak{Z} = \{(\mathbf{x}', \mathbf{y});\ \mathbf{y} \in \mathfrak{Y}_2 - \mathbf{x}'\}.$

Thus if $\mathbf{x}' \in \mathfrak{Y}_1$, $\qquad V_n[K(\mathbf{x}') \cap \mathfrak{Z}] = V_n(\mathfrak{Y}_2)$

and hence

$$V_{2n}(\mathfrak{Z}) = \int_{\mathfrak{Y}_1} V_n(K(\mathbf{x}) \cap \mathfrak{Z})\, d\mathbf{x} = V_n(\mathfrak{Y}_1) . V_n(\mathfrak{Y}_2).$$

Further $\qquad J(\mathbf{O}) \cap \mathfrak{Z} = \{(\mathbf{x}, \mathbf{O});\ \mathbf{x} \in \mathfrak{Y}_1 \text{ and } \mathbf{x} \in \mathfrak{Y}_2\},$

hence $\qquad V_n(J(\mathbf{O}) \cap \mathfrak{Z}) = V_n(\mathfrak{Y}_1 \cap \mathfrak{Y}_2).$

Let $T(\mathfrak{Z}) = \{\mathbf{y};\ (\mathbf{x}, \mathbf{y}) \in \mathfrak{Z} \text{ for some } \mathbf{x}\} = \mathfrak{Y}_2 - \mathfrak{Y}_1$, then

$$V_{2n}(\mathfrak{Z}) = \int_{T(\mathfrak{Z})} V_n(J(\mathbf{y}) \cap \mathfrak{Z})\, d\mathbf{y}.$$

Now since $\mathbf{y} \in T(\mathfrak{Z})$ there is θ, $0 \leqslant \theta \leqslant 1$ such that $\mathbf{y} \in Fr(\theta T(\mathfrak{Z}))$, that is to say $\theta^{-1}\mathbf{y} \in T(\mathfrak{Z})$. But then for some $\mathbf{x}$, $(\mathbf{x}, \theta^{-1}\mathbf{y}) \in \mathfrak{Z}$ and $\mathfrak{Z} \supset \mathfrak{H}((\mathbf{x}, \theta^{-1}\mathbf{y}), J(\mathbf{O}) \cap \mathfrak{Z})$. This last set meets $J(\mathbf{y})$ in a set congruent to $(1-\theta)(J(\mathbf{O}) \cap \mathfrak{Z})$. Thus

$$V_n(J(\mathbf{y}) \cap \mathfrak{Z}) \geqslant (1-\theta)^n V_n(\mathfrak{Y}_1 \cap \mathfrak{Y}_2).$$

Integrating over the subsets of $T(\mathfrak{Z})$ for which θ is constant,

$$V_{2n}(\mathfrak{Z}) \geqslant \int_0^1 (1-\theta)^n V_n(\mathfrak{Y}_1 \cap \mathfrak{Y}_2)\, d(\theta^n V_n(\mathfrak{Y}_2 - \mathfrak{Y}_1))$$
$$= ((n!)^2/(2n!))\, V_n(\mathfrak{Y}_1 \cap \mathfrak{Y}_2) . V_n(\mathfrak{Y}_2 - \mathfrak{Y}_1).$$

Taking $\mathfrak{Y}_1 = \mathfrak{Y}_2 = \mathfrak{X}$ we obtain the desired inequality

$$V(2\mathfrak{X}') \leqslant \binom{2n}{n} V(\mathfrak{X}).$$

The proof that this inequality is best possible is more difficult and the reader should consult the work of Rogers and Shephard [1].

CHAPTER 6

SOME SPECIAL PROBLEMS

In this chapter we consider a number of particular problems which are either of importance in themselves or which illustrate the techniques available in this branch of mathematics. The problems are extremal geometric problems; that is to say, they are inequalities stated in terms of geometrical concepts. In any particular problem it is important to define the subclass of convex sets for which the inequality becomes an equality. The problems are of a type that can, in theory at any rate, be solved by the methods of the calculus of variations. In practice these methods are difficult to apply and cumbersome to handle. In the type of problem considered here the methods given are both more elegant and more precise than those of the calculus of variations. It is possible to give only a small selection of special problems, and the actual choice may strike the reader as rather arbitrary. It is arbitrary, since to give a systematic account of a representative selection of special problems would necessitate devoting more attention to them than would be proper in an introduction to the subject.

The success of this type of method depends to a great extent upon the simplicity of the structure of the extremal figures. Where the extremal figure is unique (to within a congruence or an affine transformation say), then the method is likely to be applicable. Where the extremal figures are many and cannot easily be described in geometrical language, there the method will be difficult to apply or even impossible.

1. The isoperimetric inequality

The oldest and most important of all the extremal geometrical problems is that of finding the n-dimensional set whose surface has a given area and which contains the largest volume, or in two dimensions, that of finding the set whose boundary has given length and which covers the largest area. If this problem is considered in full generality it is necessary to define the surface area of a more or less arbitrary point set in n-dimensional space and the length of the boundary of a point set in two dimensions. We shall consider only that case in which the set is convex and we then use the definitions of area and length given in § 5·3.

The fact that of all convex sets of given volume the sphere has least surface area is an immediate consequence of Minkowski's inequality (5·5·4), in which we take $\mathfrak{X}_0 = \mathfrak{X}$ the given convex body and $\mathfrak{X}_1 = \mathfrak{S}$ the unit sphere. Then (5·5·4) becomes

$$\left[\frac{A(\mathfrak{X})}{n}\right]^n \geqslant V(\mathfrak{X})^{n-1} V(\mathfrak{S}).$$

But $A(\mathfrak{S}) = nV(\mathfrak{S})$ and thus

$$\left[\frac{A(\mathfrak{X})}{A(\mathfrak{S})}\right]^n \geqslant \left[\frac{V(\mathfrak{X})}{V(\mathfrak{S})}\right]^{n-1}.$$

Moreover, equality holds only if $\mathfrak{X}$ is homothetic to $\mathfrak{S}$, i.e. if $\mathfrak{X}$ is itself a sphere.

However, we shall give in the next theorem an alternative proof which, although considerably longer, is more illuminating. The technique used in this second method can be applied to other special problems.

Theorem 47. *Of all convex sets with a given volume in* R^n *a sphere has the least surface area.*

Let $\mathfrak{A}_r$ be the class of all convex sets $\mathfrak{X}$ for which $V(\mathfrak{X}) = V$ and which contain a sphere of radius r, $r > 0$. We shall suppose that r is so small that $\mathfrak{A}_r$ is not empty. Now write

$$D_r = \sup_{\mathfrak{X} \in \mathfrak{A}_r} D(\mathfrak{X}).$$

The number D_r is finite, since $V(\mathfrak{X}) = V$. Let $\mathfrak{B}_r$ be the subclass of those members of $\mathfrak{A}_r$ which contain a certain fixed point $\mathbf{p}$. Every member of $\mathfrak{A}_r$ is congruent to a member of $\mathfrak{B}_r$; thus if

$$\inf_{\mathfrak{Y} \in \mathfrak{B}_r} A(\mathfrak{Y}) = \inf_{\mathfrak{X} \in \mathfrak{A}_r} A(\mathfrak{X}) = A, \tag{6·1·1}$$

we may select a sequence $\mathfrak{Y}_1, \mathfrak{Y}_2, \ldots$ of members of $\mathfrak{B}_r$ such that

$$\lim_{i \to \infty} A(\mathfrak{Y}_i) = A.$$

The sets $\mathfrak{Y}_i$ all lie inside a certain bounded portion of Euclidean space. By the Blaschke selection theorem there is a subsequence of $\{\mathfrak{Y}_i\}$, say $\{\mathfrak{Y}_{i_j}\}$, which converges to a convex set $\mathfrak{Y}$. Since both the volume $V(\mathfrak{X})$ and the surface area $A(\mathfrak{X})$ are continuous functions of $\mathfrak{X}$ it follows that

$$V(\mathfrak{Y}) = V, \quad A(\mathfrak{Y}) = A. \tag{6·1·2}$$

Now if $\mathfrak{P}$ is any hyperplane there is a hyperplane $\mathfrak{Q}$ which is parallel to $\mathfrak{P}$ and about which $\mathfrak{Y}$ is symmetric. For otherwise, by property (iii) of Steiner symmetrization given in § 5·4, we should have, after symmetrization about $\mathfrak{P}$,

$$V(\mathfrak{Y}_{\mathfrak{P}}) = V, \quad A(\mathfrak{Y}_{\mathfrak{P}}) < A. \tag{6·1·3}$$

Now $\mathfrak{Y}_{\mathfrak{P}}$ certainly contains a sphere of radius r. Thus $\mathfrak{Y}_{\mathfrak{P}} \in \mathfrak{A}_r$. $\mathfrak{Y}_{\mathfrak{P}}$ may not itself belong to $\mathfrak{B}_r$ (since it may not contain $\mathbf{p}$), but a set congruent to $\mathfrak{Y}_{\mathfrak{P}}$ will belong to $\mathfrak{B}_r$ and then (6·1·3) is in contradiction with the definition of $\mathfrak{Y}$.

Select n hyperplanes that are perpendicular in pairs and about each of which $\mathfrak{Y}$ is symmetric. Then $\mathfrak{Y}$ is a central set whose centre $\mathbf{O}$ is the point of intersection of the n hyperplanes.

Next consider any hyperplane $\mathfrak{P}$ which passes through the point $\mathbf{O}$. The line through $\mathbf{O}$ perpendicular to $\mathfrak{P}$ meets $\mathfrak{Y}$ in a segment $\mathfrak{l}$ whose mid-point is $\mathbf{O}$ (since $\mathbf{O}$ is the centre of $\mathfrak{P}$). But $\mathfrak{Y}$ is symmetric about some hyperplane parallel to $\mathfrak{P}$ and must therefore be symmetric about $\mathfrak{P}$. Thus $\mathfrak{Y}$ is symmetric about every hyperplane through $\mathbf{O}$. If $\mathbf{y}_1$ and $\mathbf{y}_2$ are two points on the frontier of $\mathfrak{Y}$ the hyperplane $\mathfrak{P}$ that passes through $\mathbf{O}$ and is perpendicular to the line $\mathbf{y}_1\mathbf{y}_2$ must bisect the segment $\mathfrak{H}(\mathbf{y}_1, \mathbf{y}_2)$ (since $\mathfrak{Y}$ is symmetric about $\mathfrak{P}$) and thus the points $\mathbf{y}_1, \mathbf{y}_2$ are equidistant from $\mathbf{O}$. Hence finally $\mathfrak{Y}$ is a sphere whose centre is $\mathbf{O}$.

The theorem is proved for $\mathfrak{X} \in \mathfrak{A}_r$. But any convex set of positive volume belongs to $\mathfrak{A}_r$ for some $r > 0$ by Theorem 4, and the fact that a convex set of positive volume is necessarily n-dimensional. The theorem is completely proved.

Remarks. There are several points in this proof which are of interest.

(i) It depends upon the Blaschke selection theorem as do nearly all extremal geometrical problems.

(ii) We cannot apply the Blaschke selection theorem directly to the class of all convex sets with volume V, since the members of this class are not uniformly bounded. In order to apply the selection theorem we must make an artificial restriction and consider only those convex sets which belong to the class $\mathfrak{A}_r$. We also introduce the subclass $\mathfrak{B}_r$ of $\mathfrak{A}_r$, but the introduction of this class is natural since we are considering properties, volume and area, which are invariant under a translation. The class $\mathfrak{A}_r$ does not arise naturally in the problem; it is introduced simply for the technical reason that without it we cannot apply the Blaschke selection theorem. This situation is typical; we often need either to extend or to restrict the class with which we are dealing for purely technical reasons and then to return to the original class by some appropriate device.

(iii) At one stage in the argument (in (6·1·3)) we use the conditions under which Steiner symmetrization actually reduces the area of the convex set $\mathfrak{Y}$. It is essential to be able to use these conditions, otherwise we cannot give a description of the extremal set $\mathfrak{Y}$. In other words we need to know not only that

$$A(\mathfrak{Y}_{\mathfrak{P}}) \leqslant A(\mathfrak{Y})$$

but under what conditions we have

$$A(\mathfrak{Y}_{\mathfrak{P}}) = A(\mathfrak{Y}).$$

In this particular case it is possible to construct an argument that will prove Theorem 47 without using these conditions, but it is not as simple as that given above and it does not prove the more precise form of the theorem, namely,

Any convex set of volume V in R^n has area greater than that of a sphere of volume V unless it is actually a sphere.

This more precise form is implied by the argument given above.

This situation is also typical. It is important to know precisely the conditions under which an inequality becomes an equality. No geometrical extremal problem is solved until the class of all convex sets which possess the extreme property has been completely described.

An allied problem is that of finding the convex sets of given diameter and maximum volume. The result is that the maximum volume is attained by a sphere and only by a sphere.

By the Blaschke selection theorem we can prove that there is a set $\mathfrak{X}$ such that $D(\mathfrak{X})$ has the assigned value D and that $V(\mathfrak{X})$ is equal to the upper bound V of all the volumes of convex sets with diameter equal to D. We attempt to solve the problem by Steiner symmetrization. Symmetrization about n hyperplanes that pass through a point $\mathbf{p}$ and are perpendicular in pairs produces a set $\mathfrak{X}_1$, such that

$$D(\mathfrak{X}_1) \leqslant D, \quad V(\mathfrak{X}_1) = V,$$

and $\mathfrak{X}_1$ is central with $\mathbf{p}$ as its centre. Thus $\mathfrak{X}_1$ is contained in a sphere whose centre is $\mathbf{p}$ and whose radius is $\frac{1}{2}D$. Hence V is not greater than the volume of a sphere of radius $\frac{1}{2}D$.

To complete the solution of the problem we must show that $\mathfrak{X}$ is a sphere. Now there are no simple conditions under which the diameter of a convex set is not reduced by symmetrization, so that we cannot hope to proceed in a manner similar to that used in Theorem 47. It is clear that $\mathfrak{X}_1$ must be a sphere, and since $\mathfrak{X}_1$ is obtained from $\mathfrak{X}$ by n successive symmetrizations it would be sufficient to show that if a set $\mathfrak{X}_1$ is obtained from say $\mathfrak{X}_2$ by a symmetrization, and if $\mathfrak{X}_1$ is a sphere and $D(\mathfrak{X}_2) = D(\mathfrak{X}_1)$ then $\mathfrak{X}_2$ is a sphere. This result, though true, is difficult to prove.

We may attempt to solve the problem by a different method using central symmetrization as described in § 5·6 instead of Steiner symmetrization. Let $\mathfrak{X}$ be the set as chosen above, let $\mathfrak{X}'$ be the reflexion of $\mathfrak{X}$ in the origin and let

$$\mathfrak{X}'' = \tfrac{1}{2}(\mathfrak{X} + \mathfrak{X}').$$

Then by the Brünn-Minkowski inequality, Theorem 46,

$$V(\mathfrak{X}'') \geqslant V(\mathfrak{X}),$$

and equality holds if and only if $\mathfrak{X}$ and $\mathfrak{X}'$ are homothetic. But the choice of $\mathfrak{X}$ and the fact that $D(\mathfrak{X}'') = D(\mathfrak{X})$ imply that equality must hold. Thus $\mathfrak{X}$ is homothetic with its reflexion in the origin. But this implies that $\mathfrak{X}$ is a central set and then, as before, $\mathfrak{X}$ must be a sphere.

2. The isoperimetric inequality in R^2

In space of two dimensions it is usually possible to give much more precise information than in space of n dimensions. As an example we give first Bonnesen's refinement of the isoperimetric inequality and then Feller's solution of the second problem contained in the previous paragraph.

Let $\mathfrak{X}$ be a plane convex set whose area is S and whose boundary has length L. Then the two-dimensional form of Theorem 47 leads to the isoperimetric inequality

$$L^2 - 4\pi S \geqslant 0 \tag{6·2·1}$$

and Bonnesen's strengthened form is

$$L^2 - 4\pi S \geqslant \pi^2(R-r)^2, \tag{6·2·2}$$

where $R = R(\mathfrak{X})$ and $r = r(\mathfrak{X})$.

Consider the quadratic in x

$$f(x) = S + xL + \pi x^2. \tag{6·2·3}$$

This quadratic, which is the area of the set $\mathfrak{U}(\mathfrak{X}, x)$, has real roots if and only if (6·2·1) is correct. Thus (since $f(x) \to \infty$ as $x \to \pm\infty$) (6·2·1) will follow if we can show that there is some value of x for which $f(x) \leqslant 0$. We obtain the stronger inequality (6·2·2) by showing that $f(-r) \leqslant 0$ and $f(-R) \leqslant 0$.

We can consider Bonnesen's inequality as obtained from a particular case of an inequality involving the areas and the mixed area of two plane convex sets, one of which is inscribed in the other. We say that the plane convex set $\mathfrak{X}_2$ is inscribed in the plane convex set $\mathfrak{X}_1$ if $\mathfrak{X}_2$ is contained in $\mathfrak{X}_1$ and either there are two parallel lines which are support lines of both $\mathfrak{X}_2$ and $\mathfrak{X}_1$ or there are three lines which are support lines of both $\mathfrak{X}_2$ and $\mathfrak{X}_1$ and which bound a triangle that contains $\mathfrak{X}_2$ and $\mathfrak{X}_1$.

Then there is the following theorem:

THEOREM 48. *If $\mathfrak{X}_1, \mathfrak{X}_2$ are plane convex sets and $\mathfrak{X}_2$ is inscribed in $\mathfrak{X}_1$, then the mixed area of $\mathfrak{X}_1$ and $\mathfrak{X}_2$ is greater than or equal to the arithmetic mean of the areas of $\mathfrak{X}_1$ and $\mathfrak{X}_2$.*

This theorem is a consequence of the following lemma:

LEMMA. *In the (x, y) plane consider two polygonal arcs γ_1 and γ_2 defined by concave functions $y = f_1(x)$, $y = f_2(x)$, $-a \leqslant x \leqslant a$, where*

$$f_2(x) \leqslant f_1(x), \quad f_1(a) = f_1(-a) = f_2(a) = f_2(-a) = 0.$$

Suppose further that the segments of γ_i *are* $s_{i,j}$, $i=1,2, j=1,2,\ldots,k$, *and that* $s_{1,j}$ *is parallel to* $s_{2,j}$, $j=1,2,\ldots,k$. *Denote the perpendicular distance from a point* $\mathbf{p}$ *to* $s_{i,j}$ *by* $p_{i,j}$ *when* $\mathbf{p}$ *and the origin are both on the same side of the line containing* $s_{i,j}$. *Then*

$$\sum_{j=1}^{k} p_{2j}s_{1j} \geqslant \frac{1}{2}\sum_{j=1}^{k}(p_{1j}s_{1j}+p_{2j}s_{2j}),$$

$$\sum_{j=1}^{k} p_{1j}s_{2j} \geqslant \frac{1}{2}\sum_{j=1}^{k}(p_{1j}s_{1j}+p_{2j}s_{2j}).$$

On each segment $s_{1,j}$ we can construct a parallelogram such that one side is $s_{1,j}$, the opposite side lies in the same line as $s_{2,j}$ and the remaining sides are parallel to the y axis. This parallelogram has area $s_{1,j}(p_{1,j}-p_{2,j})$, and the set of all these parallelograms is contained in the area bounded by γ_1 and γ_2. Moreover, none of the parallelograms overlap.

Thus
$$\text{area}\,(\mathfrak{H}(\gamma_1)\dot{-}\mathfrak{H}(\gamma_2)) \geqslant \sum s_{1,j}(p_{1,j}-p_{2,j}),$$

i.e.
$$\tfrac{1}{2}\sum s_{1,j}p_{1,j}-\tfrac{1}{2}\sum s_{2,j}p_{2,j} \geqslant \sum s_{1,j}(p_{1,j}-p_{2,j}),$$

i.e.
$$\sum p_{2,j}s_{1,j} \geqslant \tfrac{1}{2}\sum(p_{1,j}s_{1,j}+p_{2,j}s_{2,j}). \tag{6·2·4}$$

Similarly, by constructing parallelograms on the segments $s_{2,j}$ with opposite sides lying in the lines $s_{1,j}$ respectively (and thus covering $\mathfrak{H}(\gamma_1)\dot{-}\mathfrak{H}(\gamma_2)$), we obtain

$$\sum s_{2,j}p_{1,j} \geqslant \tfrac{1}{2}\sum(s_{1,j}p_{1,j}+s_{2,j}p_{2,j}), \tag{6·2·5}$$

and the proof of the lemma is complete.

To prove the theorem we observe that it is sufficient to consider the case when $\mathfrak{X}_1$ and $\mathfrak{X}_2$ are polygons between whose segments there exists a 1-1 correspondence such that s_1 of Fr $\mathfrak{X}_1$ corresponds to s_2 of Fr $\mathfrak{X}_2$ if and only if s_1 is parallel to s_2 and $\mathfrak{X}_1$ lies on the same side of s_1 as does $\mathfrak{X}_2$ of s_2. For the general case can then be deduced by an approximation argument.

The frontiers of $\mathfrak{X}_1$ and $\mathfrak{X}_2$ can each be divided into two or three arcs (as the case may be) and the lemma applied to these arcs. The statement of the theorem follows immediately from the lemma.

To deduce Bonnesen's inequality, we first take $\mathfrak{X}_1$ to be $\mathfrak{X}$ and

$\mathfrak{X}_2$ to be the incircle of $\mathfrak{X}$. Then we take $\mathfrak{X}_1$ to be the circumcircle of $\mathfrak{X}$ and $\mathfrak{X}_2$ to be $\mathfrak{X}$. We obtain

$$rL \geqslant S + \pi r^2,$$

$$RL \geqslant S + \pi R^2,$$

and the inequality (6·2·2) follows.

We give next Feller's solution of the problem considered at the end of § 6·1.

Let $\mathfrak{X}$ be a plane convex set with diameter $D(\mathfrak{X})$. Select a point $\mathbf{p}$ on the frontier of $\mathfrak{X}$ as origin and introduce polar coordinates r, θ such that the interior of $\mathfrak{X}$ lies in the half-plane $0 < \theta < \pi$. Let $f(\theta)$ be the length of the radius vector in the direction θ from $\mathbf{p}$ to a point of the frontier of $\mathfrak{X}$. Then

$$V(\mathfrak{X}) = \frac{1}{2}\int_0^{\pi} [f(\theta)]^2\,d\theta = \frac{1}{2}\int_0^{\frac{1}{2}\pi} ([f(\theta)]^2 + [f(\theta+\tfrac{1}{2}\pi)]^2)\,d\theta.$$

Since any two points of $\mathfrak{X}$ are at a distance apart less than or equal to $D(\mathfrak{X})$,

$$[f(\theta)]^2 + [f(\theta+\tfrac{1}{2}\pi)]^2 \leqslant [D(\mathfrak{X})]^2 \quad (0 \leqslant \theta \leqslant \tfrac{1}{2}\pi). \tag{6·2·6}$$

Thus

$$V(\mathfrak{X}) \leqslant \pi[D(\mathfrak{X})]^2, \tag{6·2·7}$$

and equality holds in (6·2·7) only if equality holds in (6·2·6) for every θ, $0 \leqslant \theta \leqslant \frac{1}{2}\pi$. Since $\mathbf{p}$ is any point of the frontier of $\mathfrak{X}$ this means that every right-angled triangle inscribed in $\mathfrak{X}$ has a hypotenuse whose length is $D(\mathfrak{X})$. So far $\mathbf{p}$ has been an arbitrary point of the frontier of $\mathfrak{X}$; now suppose that it is a regular point so that there is exactly one support line of $\mathfrak{X}$ through $\mathbf{p}$. We know that such a point exists by the remark at the end of § 1·11. For such a choice of $\mathbf{p}$ we have

$$f(\theta) > 0 \quad (0 < \theta < \pi),$$

and therefore by (6·2·6)

$$f(\theta) < D(\mathfrak{X}) \quad (\theta \neq \tfrac{1}{2}\pi). \tag{6·2·8}$$

Let $\mathbf{q}_1, \mathbf{q}_2$ be two points of $\operatorname{Fr}\mathfrak{X}$ such that

$$\angle \mathbf{p}\mathbf{q}_1\mathbf{q}_2 = \tfrac{1}{2}\pi,$$

then if equality holds in (6·2·7) we have $|\mathbf{p}-\mathbf{q}_2|=D(\mathfrak{X})$. But this, by (6·2·8), is only possible if $\mathbf{q}_2$ lies on the radius vector perpendicular to the support line to $\mathfrak{X}$ at $\mathbf{p}$. Thus $\mathbf{q}_2$ is fixed whatever point $\mathbf{q}_1$ is taken and $\mathbf{q}_1$ lies on a circle.

The result is thus completely proved.

3. Relations between the inradius r, circumradius R, minimal width d and diameter D of a convex set

There are twelve possible inequalities between any two of r, R, d, D. Of these inequalities a number are trivial, namely,

$$r \leqslant \tfrac{1}{2}d, \quad r \leqslant R, \quad d \leqslant D, \quad D \leqslant 2R. \tag{6·3·1}$$

Since these four inequalities imply

$$d \leqslant 2R, \quad r \leqslant \tfrac{1}{2}D,$$

and since there are no inequalities of the form

$$D \leqslant \lambda r, \quad R \leqslant \lambda r, \quad R \leqslant \lambda d, \quad D \leqslant \lambda d,$$

unless we allow λ to be infinite, there are only two inequalities that are really interesting. The first gives an upper bound for R in terms of D and the second gives a lower bound for r in terms of d. We shall prove these two inequalities in this section.

THEOREM 49. *If $\mathfrak{X}$ is a convex body in R^n whose circumradius is R and whose diameter is D, then*

$$R \leqslant D[n/(2n+2)]^{\frac{1}{2}}. \tag{6·3·2}$$

It is sufficient to consider the case when $\mathfrak{X}$ is a polytope, for the general case can then be obtained by an approximation argument. Then the centre of the circumsphere is contained in the convex cover of those vertices of $\mathfrak{X}$ that lie on the frontier of the circumsphere, and therefore lies in the convex cover of s of these points where $s \leqslant n+1$ (by Carathéodory's theorem).

Take the centre of the circumsphere as origin and let the s vertices of $\mathfrak{X}$ be $\mathbf{x}_1, \ldots, \mathbf{x}_s$. Then there exist numbers $\lambda_1, \ldots, \lambda_s$ such that

$$\sum_{i=1}^{s} \lambda_i \mathbf{x}_i = 0, \quad \sum_{i=1}^{s} \lambda_i = 1,$$

where $\lambda_i \geqslant 0$, $1 \leqslant i \leqslant s$. Let D_1 be the maximum of $|\mathbf{x}_i - \mathbf{x}_j|$ for any i, j. For any j, $1 \leqslant j \leqslant s$,

$$(1-\lambda_j) D_1^2 = \sum_{i=1}^{s} \lambda_i D_1^2 - \lambda_j D_1^2 \geqslant \sum_{i=1}^{s} \lambda_i |\mathbf{x}_i - \mathbf{x}_j|^2$$

$$= \sum_{i=1}^{s} \lambda_i (2R^2 - 2\mathbf{x}_i . \mathbf{x}_j) = 2R^2.$$

Summing this inequality for j from 1 to s we obtain

$$(s-1) D_1^2 \geqslant 2sR^2.$$

Thus $$D \geqslant D_1 \geqslant R\sqrt{\frac{2s}{s-1}} \geqslant R\sqrt{\frac{2(n+1)}{n}},$$

and this is the required result.

We next prove the 'dual' result.

THEOREM 50. *If $\mathfrak{X}$ is of minimal width d then r, the inradius, satisfies*

$$r \geqslant d/(2n^{\frac{1}{2}}) \quad n \text{ odd} \tag{6·3·3}$$

$$r \geqslant d(n+2)^{\frac{1}{2}}/2(n+1) \quad n \text{ even}. \tag{6·3·4}$$

It is sufficient to prove this result when $\mathfrak{X}$ is such that its frontier contains $n+1$ points $\mathbf{x}_0, \ldots, \mathbf{x}_n$ which lie on the insphere and which form the vertices of a simplex of which the incentre is an interior point. For if this is not true and if $\epsilon > 0$ is given we can find $n+1$ points $\mathbf{z}_0, \ldots, \mathbf{z}_n$ on the frontier of the insphere, distributed in the manner described and such that if the closed half-space bounded by the hyperplanes tangent to the insphere at $\mathbf{z}_i$ and containing the insphere is $\mathfrak{P}_i$, then the set

$$\mathfrak{X}_1 = \mathfrak{X} \cap \mathfrak{P}_0 \cap \mathfrak{P}_1 \cap \ldots \cap \mathfrak{P}_n$$

has minimal width of at least $d - \epsilon$. The appropriate inequality is true for $\mathfrak{X}_1$, and therefore, in the limit as $\epsilon \to 0$, it is true also for $\mathfrak{X}$.

We assume then that the points $\mathbf{x}_0, \mathbf{x}_1, \ldots, \mathbf{x}_n$ belong to $\mathfrak{X}$ as described. The hyperplanes tangent to the insphere at these points bound a simplex $\mathfrak{Y}$ and

$$d(\mathfrak{Y}) \geqslant d(\mathfrak{X}) = d.$$

Thus it is sufficient to prove (6·3·3) or (6·3·4) for $\mathfrak{Y}$.

Denote the vertices of $\mathfrak{Y}$ by $\mathbf{y}_1, \mathbf{y}_2, \ldots, \mathbf{y}_{n+1}$. We call that face of $\mathfrak{Y}$ that does not contain $\mathbf{y}_j$, the jth face of $\mathfrak{Y}$. Denote the vector which is perpendicular to the jth face of $\mathfrak{Y}$, whose magnitude is equal to the $(n-1)$-dimensional volume of the jth face and which points outward from $\mathfrak{Y}$ by $\mathbf{v}_j$.

The vector $\mathbf{v}=\mathbf{v}_1+\mathbf{v}_2+\ldots+\mathbf{v}_{n+1}$ is such that if $\mathbf{u}$ is a unit vector then $\mathbf{v}.\mathbf{u}$ is equal to the sum of the signed areas of the projections of the faces of $\mathfrak{Y}$ in the direction $\mathbf{u}$. Since $\mathfrak{Y}$ is a closed solid $\mathbf{v}.\mathbf{u}=0$. Thus as $\mathbf{u}$ is any unit vector,

$$\mathbf{v}_1+\mathbf{v}_2+\ldots+\mathbf{v}_{n+1}=\mathbf{O}. \tag{6·3·5}$$

Squaring gives

$$\Sigma\mathbf{v}_j^2+2\sum_{i<j}\mathbf{v}_i.\mathbf{v}_j=0. \tag{6·3·6}$$

Let $\mathfrak{S}(k)$ be the class of all sets of k distinct integers selected from $1,2,\ldots,n+1$. Let $s(k)$ be a typical member of $\mathfrak{S}(k)$ and let $s'(k)$ be the set of $n+1-k$ integers that belong to $1,2,\ldots,n+1$ and not to $s(k)$.

Suppose now that k is fixed and write $\mathfrak{S}, s, s'$ instead of $\mathfrak{S}(k), s(k), s'(k)$. Write

$$\mathbf{v}(s)=\sum_{j\in s}\mathbf{v}_j.$$

This vector is perpendicular to any vector that is perpendicular to $\mathbf{v}_j, j\in s$. Thus it is perpendicular to the space spanned by the vertices $\mathbf{y}_j, j\in s'$. But by (6·3·5) $\mathbf{v}(s)$ can also be expressed as

$$\mathbf{v}(s)=-\sum_{j\in s'}\mathbf{v}_j,$$

and so it is perpendicular to the space spanned by $\mathbf{y}_j, j\in s$. Hence it is perpendicular to the two parallel support hyperplanes of $\mathfrak{Y}$, one of which contains $\mathbf{y}_j, j\in s$ and the other $\mathbf{y}_j, j\in s'$. Denote the distance apart of these two hyperplanes by $d(s)$.

$|\mathbf{v}(s)|$ is the sum of signed areas of projections of faces of $\mathfrak{Y}$ in the direction of $\mathbf{v}(s)$, and it can be seen, by subdividing the simplex, that

$$|\mathbf{v}(s)|\,d(s)=nV(\mathfrak{Y}), \tag{6·3·7}$$

where $V(\mathfrak{Y})$ is also equal to $\frac{1}{n}r\Sigma|\mathbf{v}_j|$. Thus d satisfies

$$d\leqslant\min d(s)\leqslant r\min_{1\leqslant k\leqslant n}\min_{s\in\mathfrak{S}(k)}(\Sigma|\mathbf{v}_j|)/|\mathbf{v}(s)|. \tag{6·3·8}$$

Now

$$\max_{s\in\mathfrak{S}(k)}|\mathbf{v}(s)|\geqslant\left\{\sum_{s\in\mathfrak{S}(k)}[\mathbf{v}(s)]^2\Big/\binom{n+1}{k}\right\}^{\frac{1}{2}}$$

$$=\left\{\binom{n-1}{k-1}\Sigma\mathbf{v}_j^2\Big/\binom{n+1}{k}\right\}^{\frac{1}{2}}, \tag{6·3·9}$$

from (6·3·6). Also

$$\Sigma\,|\,\mathbf{v}_j\,| \leqslant \{\Sigma\mathbf{v}_j^2 + 2\sum_{i<j}|\,\mathbf{v}_j\,|\,.\,|\,\mathbf{v}_i\,|\}^{\frac{1}{2}} \leqslant \{(n+1)\,\Sigma\mathbf{v}_j^2\}^{\frac{1}{2}}, \quad (6\cdot3\cdot10)$$

where the last inequality is easily established by induction since $|\,\mathbf{v}_i\,|$ are positive numbers. Thus substituting from (6·3·9) and (6·3·10) in (6·3·8) we have

$$d \leqslant r \min_{1\leqslant k\leqslant n}\left\{\frac{(n+1)^2 n}{k(n+1-k)}\right\}^{\frac{1}{2}}. \quad (6\cdot3\cdot11)$$

As k varies through integral values between 1 and n the least value of the left-hand side of (6·3·11) occurs when $k=\frac{1}{2}n$ if n is even or when $k=\frac{1}{2}(n+1)$ if n is odd. The corresponding inequalities that result on substituting these values in (6·3·11) are precisely the inequalities (6·3·3) and (6·3·4).

Remarks. It is interesting to compare Theorems 49 and 50. The results are dual results, but the difficulty of proving Theorem 50 is much greater than that of proving Theorem 49. Theorem 49 was proved over fifty years ago, and since then has been re-proved and re-discovered many times. Theorem 50 has been proved exactly once some thirty years ago, and the proof given here is essentially a geometrical form of that argument which was based on the properties of determinants and cofactors of determinants.

The proofs of both theorems begin by a simplification. This is a standard procedure, as one nearly always needs to clear away some irrelevancy. Since the extremal figures in each case will include regular simplexes it is natural to consider polytopes and then obtain the general result by an approximation argument. There is some choice of method here, and the two theorems illustrate two different methods of procedure. The first depends upon the Blaschke selection theorem, the second does not.

The succeeding arguments are quite different. In the first we use a geometrical analogue of the calculus of variations procedure to consider variations in the circumradius due to variations in the positions of the vertices of the simplex but make no attempt to calculate the circumradius of a general simplex. In the second the procedure is essentially to calculate the width in a variety of directions and then to pick the least of these widths.

It is possible to construct arguments which will cover both cases if we consider analogous problems on the surface of a sphere and then consider the limit as the radius of the sphere tends to infinity. See the paper by Santaló referred to in the notes.

4. Plane convex sets

It is often possible to investigate problems on plane convex sets much more easily and more thoroughly than their analogues in a space of higher dimensions. There are many problems that have been solved in two-dimensional space and whose solution in three dimensions is still unknown. A number of these problems will be mentioned in this section.

It will be assumed throughout this paragraph that all the convex sets to be used have interior points.

Central convex sets and asymmetry

The definition of a central convex set was given in § 4·5: it is a convex set $\mathfrak{X}$ whose reflexion in a point $\mathbf{p}$ coincides with $\mathfrak{X}$. The point $\mathbf{p}$ is uniquely defined by this property and is called the centre of $\mathfrak{X}$. Another way of stating the definition is that every line through $\mathbf{p}$ meets $\mathfrak{X}$ in a segment of which $\mathbf{p}$ is the mid-point. Stated in this manner it is clear that any affine transformation transforms central convex sets into central convex sets.

Suppose now that we consider a property of convex sets which is true for all central convex sets and is not true for all convex sets. Then we shall naturally be interested in those convex sets for which this property is 'most untrue', where of course the meaning of 'most untrue' has to be appropriately defined for each particular problem. It is the case that the class of convex sets which may most naturally be considered as antithetic to the central convex sets are triangles. We shall consider three properties for which this is true.

(i) *Besicovitch's definition of asymmetry*

One way in which we can measure the asymmetry of a convex set $\mathfrak{X}$ is to find how closely we can approximate to $\mathfrak{X}$ by a central convex set, say $\mathfrak{Y}$. We may impose other restrictions on $\mathfrak{Y}$. For example, we may insist that $\mathfrak{Y} \subset \mathfrak{X}$ or $\mathfrak{Y} \supset \mathfrak{X}$. We give here a result due to Besicovitch.

THEOREM 51. *Every plane convex set $\mathfrak{X}$ of area $V(\mathfrak{X})$ contains a central convex set $\mathfrak{Y}$ such that*

$$V(\mathfrak{Y}) \geqslant \tfrac{2}{3}V(\mathfrak{X}).$$

We can find a point $\mathbf{p}$ such that there are three lines that pass through $\mathbf{p}$ and divide the plane into six sectors, each of which contains a subset of $\mathfrak{X}$ of area $\frac{1}{6}V(\mathfrak{X})$. (Whether we consider the sectors as open or closed is immaterial.) For in every direction θ, $0 \leqslant \theta < 2\pi$, there is a line $l(\theta)$ that divides $\mathfrak{X}$ into two sets of equal area. If $|\theta - \phi| = \pi$, then $l(\theta)$ coincides with $l(\phi)$. Given θ there are two distinct directions $\alpha_1(\theta)$ and $\alpha_2(\theta)$ such that $l(\theta)$ and $l(\alpha_i(\theta))$ divide the plane into four sectors of which two contain parts of $\mathfrak{X}$ of area $\frac{1}{6}V(\mathfrak{X})$ and two contain parts of $\mathfrak{X}$ of area $\frac{1}{3}V(\mathfrak{X})$. If $l(\theta), l(\alpha_1(\theta))$ and $l(\alpha_2(\theta))$ are concurrent their point of concurrence is the point $\mathbf{p}$ and the statement is proved. Otherwise, let the point at which $l(\theta)$ meets $\operatorname{Fr}\mathfrak{X}$ and is in the sense outward from $\mathfrak{X}$ be $\mathbf{x}(\theta)$. Assign coordinates at $\mathbf{x}(\theta)$, x in the direction of $l(\theta)$ and y perpendicular to $l(\theta)$ in the counter-clockwise sense round $\operatorname{Fr}\mathfrak{X}$. Denote the y coordinate of the point of intersection of $l(\alpha_1(\theta))$ and $l(\alpha_2(\theta))$ by $\eta(\theta)$. As θ varies $\eta(\theta)$ varies continuously and if $|\theta - \phi| = \pi$ then $\eta(\theta) = -\eta(\phi)$, because although the point of intersection is the same in both cases the direction in which its distance from $l(\theta)$ is measured has been reversed. Thus for some $\theta, \eta(\theta) = 0$, and we have the three concurrent lines as required.

We shall call $\mathbf{p}$ a six-partite point and the three lines through $\mathbf{p}$ we shall call division lines. By an appropriate affine transformation we may reduce the general case to the particular case in which the division lines make angles of 60° with one another In what follows we shall consider this case only.

Let the division lines be $\mathbf{apa'}$, $\mathbf{bpb'}$, $\mathbf{cpc'}$ meeting the frontier of $\mathfrak{X}$ in $\mathbf{aa'}$, $\mathbf{bb'}$, $\mathbf{cc'}$, and suppose that the points $\mathbf{abca'b'c'}$ are in anticlockwise order round the frontier of $\mathfrak{X}$ (see fig. 12). On $\mathbf{pa}$, $\mathbf{pb}$, produced if necessary, let $\mathbf{x}, \mathbf{y}, \mathbf{x'}, \mathbf{y'}$ be four points such that $\mathbf{xx'}$ lie on $\mathbf{pa}$ and $\mathbf{yy'}$ lie on $\mathbf{pb}$; $\mathbf{xy}$ is perpendicular to $\mathbf{pb}$, $\mathbf{x'y'}$ is perpendicular to $\mathbf{pa}$; the area of the triangle $\mathbf{pxy}$ equals that of $\mathbf{px'y'}$ and both are equal to $\frac{1}{6}V(\mathfrak{X})$. Let $\mathbf{z}$ be the point of intersection of $\mathbf{xy}$ and $\mathbf{x'y'}$.

We show next that the point **z** is contained in $\mathfrak{X}$. Suppose that a line through **z** cuts **pa** in **s** and **pb** in **t**. The area of the triangle **pst** is less than that of **pxy** unless either **s** lies in segment **px**′ or **t** in segment **py**. Suppose now that **z** was exterior to $\mathfrak{X}$ and that the line **szt** did not meet $\mathfrak{X}$. The part of $\mathfrak{X}$ in the sector bounded by the half-lines **pa** and **pb** produced is contained in the triangle **stp** and has area less than that of this triangle. By the remark made above, **st** must meet an interior point either of the segment **px**′ or of the segment **py**. Now suppose that the first alternative

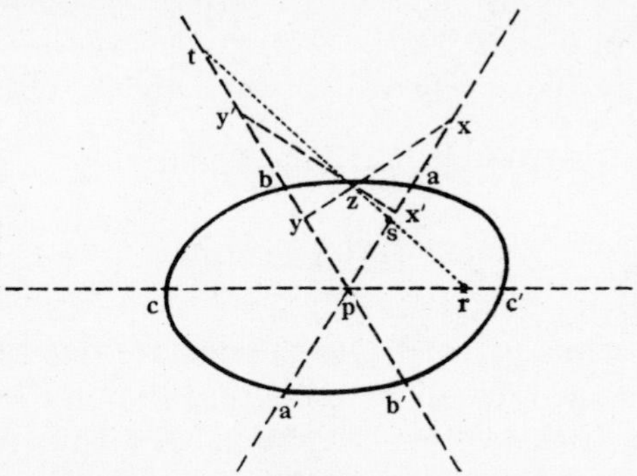

Fig. 12

held and **st** met **px**′ in **s** and **pc**′ in **r**. Then the area of the triangle **psr** is less than that of the triangle **px**′**y**′, i.e. less than $\frac{1}{6}V(\mathfrak{X})$. But this is impossible because the triangle **psr** contains that part of $\mathfrak{X}$ in the sector **apc**′ and this is of area $\frac{1}{6}V(\mathfrak{X})$. Similarly, **st** cannot meet an interior point of the segment **py**. Thus every line through **z** meets $\mathfrak{X}$ and hence $\mathbf{z}\in\mathfrak{X}$.

By an exactly similar argument the five points obtained from **z** by reflexion in the lines **apa**′, **bpb**′, **cpc**′ also belong to $\mathfrak{X}$. These five points, together with **z**, form the vertices of a regular hexagon of area $\frac{2}{3}V(\mathfrak{X})$. Since this hexagon is contained in $\mathfrak{X}$ Besicovitch's result is proved.

Thus for any convex set $\mathfrak{X}$ the number λ defined by

$$\lambda = \max_{\mathfrak{Y}} V(\mathfrak{Y})/V(\mathfrak{X}),$$

where the maximum is taken over the class of all central convex sets contained in $\mathfrak{X}$, varies between 1 and $\frac{2}{3}$. The upper bound 1 is attained if and only if $\mathfrak{X}$ is central, whilst the lower bound is attained if and only if $\mathfrak{X}$ is a triangle. (This last statement is proved by considering the possible cases of equality in the argument of Theorem 51.)

(ii) *Division of a convex set by lines through a point*

For a central convex set $\mathfrak{X}$ there is a point **p**, namely, the centre of $\mathfrak{X}$, such that every line through **p** divides $\mathfrak{X}$ into two sets of equal area. For a convex set that is not central no such point exists. Denote by $l(\mathbf{p};\, \theta)$ the line through **p** in the direction θ and by $M(\mathbf{p};\, \theta)$, $m(\mathbf{p};\, \theta)$ the areas of the two sets into which $\mathfrak{X}$ is divided by $l(\mathbf{p};\, \theta)$, where $M(\mathbf{p};\, \theta) \geqslant m(\mathbf{p};\, \theta)$. Write

$$\mu(\mathbf{p}) = \underset{\theta}{\text{up. bd.}}\,[M(\mathbf{p};\, \theta)/m(\mathbf{p};\, \theta)],$$

$$\mu = \mu(\mathfrak{X}) = \min_{\mathbf{p}} \mu(\mathbf{p}).$$

Now $\mu = 1$ if and only if $\mathfrak{X}$ is central and the question arises as to the maximum value of μ. Actually $\mu \leqslant 1\frac{1}{4}$ and $\mu = 1\frac{1}{4}$ if and only if $\mathfrak{X}$ is a triangle. This last statement follows from (i), for μ is an affine invariant, and with the notation used in (i) if $l(\mathbf{p};\, \theta)$ lies inside the two sectors **apa′** and **bpb′**, then the part of $\mathfrak{X}$ lying inside these sectors and on one side of $l(\mathbf{p};\, \theta)$ has area at least equal to that of the regular hexagon in the same regions together with $\frac{1}{3}V(\mathfrak{X})$ which lies outside these sectors. Thus

$$m(\mathbf{p};\, \theta) \geqslant \tfrac{4}{9}V(\mathfrak{X}), \quad \mu(\mathbf{p}) \leqslant \tfrac{5}{4}.$$

Equality can hold only if the part of $\mathfrak{X}$ in the sector **apb** is bounded by either **xy** or **x′y′** and the part in the sector **a′pb′** by the reflexion of **x′y′** or **xy** respectively in **p**. From the restrictions on the areas of the parts of $\mathfrak{X}$ in the other sectors it then follows that $\mathfrak{X}$ is a triangle.

(iii) *Division of a convex set by three lines, not necessarily concurrent*

Consider a convex set $\mathfrak{X}$ which is divided by three non-concurrent straight lines into six regions of equal area and one other triangular region. It is not possible to divide every convex set in this manner, for example, if $\mathfrak{X}$ is central, unless we allow concurrent lines of division and consider a point to be a degenerate triangle. In this case the triangle is of zero area, and it is natural to inquire as to the upper bound of the triangular area in relation to the value of the six equal areas. We shall show here that the area of this triangle is at most $\frac{1}{8}$ that of each of the other six regions.

It is assumed that the triangular region is an equilateral triangle. This does not involve any real loss of generality, for, if

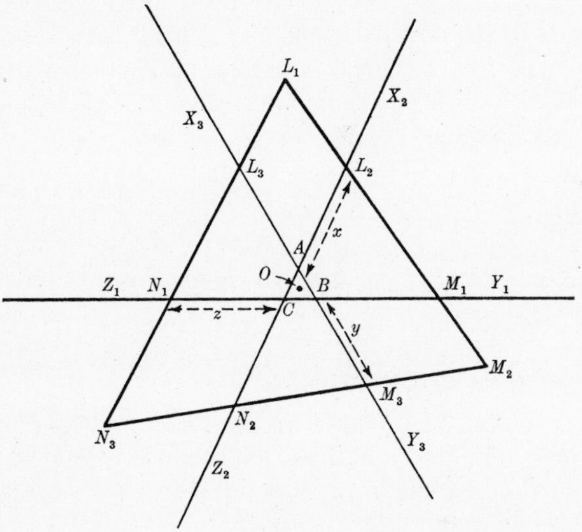

Fig. 13

it were not true there would be an affine transformation of the plane which transforms the triangle into an equilateral triangle. We could then apply the following argument to the transform of $\mathfrak{X}$.

Denote the triangular region by ABC where the order ABC is clockwise, let μ denote the area of the triangle ABC and let O be the centroid of ABC (see fig. 13). Let AB, BC, CA produced in both directions meet $\text{Fr}\,\mathfrak{X}$ in X_3Y_3, Y_1Z_1, X_2Z_2, respectively, where the order of these points in the counter-clockwise direction round $\text{Fr}\,\mathfrak{X}$ is $X_3Z_1Z_2Y_3Y_1X_2$ and where the order of the points on AB is X_3ABY_3.

Draw a line L_2M_1 parallel to X_2Y_1 meeting AX_2 in L_2 and BY_1 in M_1 so that the area of the quadrilateral L_2ABM_1 is equal to the area of the region bounded by $\mathrm{Fr}\,\mathfrak{X}$, AX_2, AB and BY_1. Denote this common area by λ. Similarly, draw L_3N_1 parallel to X_3Z_1 and M_3N_2 parallel to Z_2Y_3, where L_3 is on AB, N_1 on BC, N_2 on AC and M_3 on AB, so that

$$\text{area } L_3N_1CA = \text{area } M_3N_2CB = \lambda.$$

Let L_1, M_2, N_3 be the points of intersection of L_2M_1 and L_3N_1, M_3N_2 and M_1L_2, M_3N_2 and L_3N_1, respectively.

The part of the domain inside $\mathfrak{X}$ contained between BY_1 and BY_3 is by hypothesis of area λ and is contained in the quadrilateral $BM_1M_2M_3$. From this and similar considerations it follows that

$$\lambda \leqslant \min\{\text{area } BM_1M_2M_3, \text{area } AL_1L_2L_3, \text{area } CN_1N_3N_2\}. \quad (6{\cdot}4{\cdot}1)$$

Denote the minimum on the right-hand side of (6·4·1) by $f(L_1M_2N_3)$.

Next consider the triangle $L_1M_2N_3$ as a member of that class of triangles which are such that the areas ABM_1L_2, ACN_1L_3 and BCN_2M_3 are all equal to λ. The maximum of $f(L_1M_2N_3)$ as $L_1M_2N_3$ varies in this class occurs when the areas $BM_1M_2M_3$, $AL_2L_3L_1$ and $CN_1N_3N_2$ are all equal.

Write x, y, z for the lengths respectively of AL_2, BM_3, CN_1. These three numbers are three independent variables which determine the triangle $L_1M_2N_3$ uniquely in the class defined above. Also we may write

$$\text{area } BM_1M_2M_3 = F(x,y); \quad \text{area } CN_2N_3N_1 = F(y,z);$$
$$\text{area } AL_3L_1L_2 = F(z,x).$$

The following lemma is required.

LEMMA. *Let $F(x,y)$ be a function of two variables which is strictly decreasing in x for every fixed y and strictly increasing in y for every fixed x, then if*

$$F(x,y) = F(y,z) = F(z,x),$$

it follows that $$x = y = z.$$

The proof of the lemma is obvious.

Thus the maximum of $f(L_1M_2N_3)$ occurs where $x = y = z$. In this case the triangle $L_1M_2N_3$ is equilateral with O as its centroid.

But $f(L_1M_2N_3)$ is a maximum, say F, only when the area of the triangle $L_1M_2N_3$ is a maximum, and, since $L_1M_2N_3$ is equilateral, this occurs when the distance from O to L_1M_2 is a maximum. But the class of triangles to which $L_1M_2N_3$ belongs is such that as $L_1M_2N_3$ varies in this class L_1M_2 envelopes a hyperbola and the maximal distance from O to L_1M_2 occurs when L_1M_2 is parallel to AB. Thus the equilateral triangle of largest area, $L_1M_2N_3$, is homothetic to ABC.

Hence from (6·4·1) $\lambda \leqslant F$, and direct calculation of F shows that this implies that $\mu \leqslant \frac{1}{8}\lambda$.

Remark. A similar method may be applied in the case of subdivision into seven areas by three non-concurrent straight lines such that three of the areas have one common value and three others another common value.

As regards the three-dimensional problems, analogous to (i), (ii), (iii), it is natural to conjecture as a solution of (i) that of all three-dimensional convex sets a tetrahedron is the most asymmetrical, but whether this is so or not is still unknown. The solution to (ii) can be easily obtained (see B. H. Neumann [1] in this connection) but not by a method analogous to that given here. For (iii) there appears to be no easily recognizable analogue.

CHAPTER 7

SETS OF CONSTANT WIDTH

A set of constant width is a bounded closed convex set for which every two parallel support hyperplanes are at the same distance apart. It is at first rather surprising that there should exist any such sets, apart from spheres, but in fact there is a great variety of these sets, although they do have certain properties in common with spheres.

A fundamental property of sets of constant width is that they are complete,† that is to say, if $\mathfrak{X}$ is of constant width and $\mathfrak{Y} \supset \mathfrak{X}$, then either $\mathfrak{Y} = \mathfrak{X}$ or $D(\mathfrak{Y}) > D(\mathfrak{X})$. Or we may say that if $\mathfrak{X}$ is of constant width then the addition of any point to $\mathfrak{X}$ increases its diameter. The converse is also true, a complete set is a set of constant width.

There are five points of view from which we may regard sets of constant width and which lead to five types of properties.

(i) We can consider the direct consequences of the definition. The result of J. Pàl that every set of constant width in R^2 can be inscribed in a regular hexagon (every side of the hexagon actually touching the set) is a result of this type.

(ii) We can regard the class of sets of constant width as a generalization of the class of spheres and consider what properties of spheres remain true for all sets of constant width, and to what extent the properties of spheres are not true for sets of constant width. For example, the insphere and circumsphere of a set of constant width are concentric (as they are for a sphere!); every plane convex set of minimal width d contains a circle of diameter $d/3^{\frac{1}{2}}$ (Theorem 50), and it is natural to inquire what is the largest number $\lambda = \lambda(d)$ such that every plane convex set of minimal width d contains a set of constant width λ.

(iii) If $\mathfrak{X}$ is a set of constant width λ any circle of radius λ meets $\mathfrak{X}$ in a single arc (or a point or the empty set). This corresponds to the property of a convex set that any line meets it in a single

† This should not be confused with the topological use of the same word.

segment (or a point or the empty set). Thus sets of constant width are to the class of circles of radius λ as convex sets are to straight lines. A property which follows is that through every point of Fr $\mathfrak{X}$ there passes a circle of radius λ that bounds a disk containing $\mathfrak{X}$, i.e. a supporting circle.

(iv) We can consider any geometrical extremal property of convex sets restricted to the class of sets of constant width. For example, we could consider which sets of constant width are 'most' asymmetrical in the sense of Besicovitch (see 6·4 (i) and 7·2 (vi)).

(v) Finally, the fact that convex sets are complete may be used to obtain information about the class of all convex sets or indeed about point sets generally. For example, in proving that every plane set of diameter D can be divided into three sets each of diameter less than D, we need only consider sets of constant width.

It will be seen from the above that there is a great deal of scope for research into the structure of sets of constant width. A factor which makes such investigations rather difficult is that the property of being of constant width is not affine invariant, nor is it invariant under Steiner symmetrization. However, it is invariant under vector addition (i.e. the vector sum of two sets of constant width is itself of constant width).

1. General properties

We prove first that the class of sets of constant width λ is identical with the class of complete sets of diameter λ. An auxiliary property is established which is described in terms of the following notation. If $\mathfrak{X}$ is a convex set and $D(\mathfrak{X}) = \lambda$, we define $\mathfrak{S}(\mathfrak{X})$ to be the intersection of all spheres of radius λ whose centres belong to $\mathfrak{X}$. $\mathfrak{S}(\mathfrak{X})$ is a closed convex set and $\mathfrak{S}(\mathfrak{X}) \supset \mathfrak{X}$. Then we have the following theorem:

THEOREM 52. *The following two statements are equivalent:*

(i) *$\mathfrak{X}$ is a set of constant width λ.*

(ii) *$\mathfrak{X}$ is a complete set of diameter λ.*

Further, if (i) *or* (ii) *is true then*

(iii) *$\mathfrak{X}$ is a convex set of diameter λ and $\mathfrak{S}(\mathfrak{X}) = \mathfrak{X}$.*

We shall prove the theorem by establishing the implications

$$(\text{i}) \Rightarrow (\text{ii}), \quad (\text{ii}) \Rightarrow (\text{iii}), \quad (\text{ii}) \Rightarrow (\text{i}).$$

Proof that (i)$\Rightarrow$(ii). First, $\mathfrak{X}$ is of diameter λ because there are two points $\mathbf{x}_1, \mathbf{x}_2$ of $\mathfrak{X}$ such that $|\mathbf{x}_1 - \mathbf{x}_2| = D(\mathfrak{X})$; the hyperplanes through $\mathbf{x}_1$ and $\mathbf{x}_2$ perpendicular to $\mathbf{x}_1\mathbf{x}_2$ are support hyperplanes of $\mathfrak{X}$ and thus, since the width apart of these hyperplanes is $D(\mathfrak{X})$, we have $D(\mathfrak{X}) = \lambda$. Let $\mathbf{y}$ be a point exterior to $\mathfrak{X}$ and let $\mathbf{x}$ be the point of $\mathfrak{X}$ nearest to $\mathbf{y}$. The hyperplane through $\mathbf{x}$ perpendicular to $\mathbf{xy}$ is a support hyperplane of $\mathfrak{X}$. The parallel support hyperplane to $\mathfrak{X}$ meets $\mathfrak{X}$ in at least one point, say $\mathbf{z}$. Further, we may suppose that $\mathbf{z}$ lies on $\mathbf{yx}$ produced because otherwise we should have $D(\mathfrak{X}) > \lambda$. Then

$$|\mathbf{y} - \mathbf{z}| > |\mathbf{x} - \mathbf{z}| = \lambda,$$

and $\mathbf{y}$ cannot be added to $\mathfrak{X}$ without increasing its diameter. Thus $\mathfrak{X}$ is a complete set of diameter λ.

Proof that (ii)$\Rightarrow$(iii). If $\mathfrak{X}$ is a complete set, then since, by Theorem 12, $D(\mathfrak{H}(\mathfrak{X})) = D(\mathfrak{X})$, it follows that $\mathfrak{H}(\mathfrak{X}) = \mathfrak{X}$ and thus $\mathfrak{X}$ is convex. We have to show that $\mathfrak{X} \supset \mathfrak{S}(\mathfrak{X})$. If this is not so there is a point $\mathbf{y}$ that belongs to $\mathfrak{S}(\mathfrak{X})$ and not to $\mathfrak{X}$. But then the set formed by the union of $\mathbf{y}$ with $\mathfrak{X}$ is of diameter $D(\mathfrak{X})$, thus $\mathfrak{X}$ is not complete. This contradiction shows that (ii)$\Rightarrow$(iii).

Proof that (ii)$\Rightarrow$(i). Suppose that this implication is not true and that there exists a set $\mathfrak{X}$ that satisfies (ii) and is of minimal width d, $d < \lambda$. Let $\mathfrak{P}_1$ and $\mathfrak{P}_2$ be two parallel support hyperplanes of $\mathfrak{X}$ that are at a distance d apart. There are points $\mathbf{x}_1, \mathbf{x}_2$ of $\mathfrak{X}$ on $\mathfrak{P}_1, \mathfrak{P}_2$ respectively, such that $\mathbf{x}_1\mathbf{x}_2$ is perpendicular to $\mathfrak{P}_1$ and to $\mathfrak{P}_2$. Thus

$$|\mathbf{x}_1 - \mathbf{x}_2| = d < \lambda.$$

Now there exists a point $\mathbf{q}$ of $\mathfrak{X}$ such that $|\mathbf{q} - \mathbf{x}_1| = \lambda$. By (iii), $\mathfrak{S}(\mathfrak{X}) = \mathfrak{X}$, and since $\mathfrak{S}(\mathfrak{X})$ is an intersection of spheres of radius λ it follows that if $\mathbf{x}_1, \mathbf{x}_2 \in \mathfrak{X}$ then so does any circular arc of radius λ with end-points $\mathbf{x}_1, \mathbf{x}_2$. In the plane $\mathbf{x}_1\mathbf{x}_2\mathbf{q}$ there is an arc of radius λ with end-points $\mathbf{x}_2$ and $\mathbf{q}$, and which crosses the hyperplane $\mathfrak{P}_2$. Since $\mathfrak{P}_2$ is a support hyperplane of $\mathfrak{X}$ we have a contradiction. Hence the assumption that (i) is false is itself false and (ii)$\Rightarrow$(i).

Remark. Every set of constant width in R^n contains interior points and, from property (iii), every support hyperplane to a set of constant width is a regular support hyperplane.

THEOREM 53. *The insphere and the circumsphere of a set of constant width λ are concentric and the sum of their radii is equal to λ.*

As before, let $\mathfrak{J}(\mathfrak{X})$ and $\mathfrak{C}(\mathfrak{X})$ be the insphere and the circumsphere of the set $\mathfrak{X}$ of constant width λ. Let $\mathbf{p}$ be the centre of $\mathfrak{J}(\mathfrak{X})$ and $r(\mathfrak{X})$ be its radius. Then $\operatorname{Fr}(\mathfrak{J}(\mathfrak{X}))$ meets the frontier of $\mathfrak{X}$ and in fact the point $\mathbf{p}$ is contained in the convex cover of $(\operatorname{Fr}\mathfrak{J}(\mathfrak{X}))\cap\operatorname{Fr}\mathfrak{X}$. Thus there are k points of this set, say $\mathbf{x}_1, \mathbf{x}_2, \ldots, \mathbf{x}_k$, with $k \leqslant n+1$, such that $\mathbf{p}$ belongs to the simplex $\mathfrak{H}(\mathbf{x}_1, \mathbf{x}_2, \ldots, \mathbf{x}_k)$. The hyperplane through $\mathbf{x}_j$ perpendicular to $\mathbf{px}_j$ is the only hyperplane through $\mathbf{x}_j$ that does not cut $\mathfrak{J}(\mathfrak{X})$, and thus, since $\mathfrak{X} \supset \mathfrak{J}(\mathfrak{X})$, it is the only hyperplane through $\mathbf{x}_j$ that does not cut $\mathfrak{X}$. Hence this hyperplane is the support hyperplane to $\mathfrak{X}$ at $\mathbf{x}_j$. Let $\mathbf{y}_j$ be the point on $\mathbf{x}_j\mathbf{p}$ produced so that $|\mathbf{x}_j - \mathbf{y}_j| = \lambda$ and $\mathbf{p}$ lies between $\mathbf{x}_j$ and $\mathbf{y}_j$. Each of the points $\mathbf{y}_j$ is at the same distance from $\mathbf{p}$. Denote the sphere whose centre is $\mathbf{p}$ and which contains these points in its frontier by $\mathfrak{S}$. Now $\mathfrak{S} \supset \mathfrak{X}$, for if $\mathbf{q}$ is a point of $\mathfrak{X} \dot{-} \mathfrak{S}$, $\mathbf{qp}$ produced meets $\operatorname{Fr}(\mathfrak{J}(\mathfrak{X}))$ in (say) the point $\mathbf{t}$. Since $\mathfrak{X} \supset \mathfrak{J}(\mathfrak{X})$, $\mathbf{t} \in \mathfrak{X}$. Thus

$$D(\mathfrak{X}) \geqslant |\mathbf{q}-\mathbf{t}| = |\mathbf{q}-\mathbf{p}| + |\mathbf{p}-\mathbf{t}| > |\mathbf{y}_j-\mathbf{p}| + |\mathbf{p}-\mathbf{x}_j| = \lambda.$$

This is impossible as the diameter of $\mathfrak{X}$ is its width. Thus $\mathfrak{S} \supset \mathfrak{X}$. Also $\mathbf{y}_j \in \mathfrak{X}$. But $\mathbf{p}$ belongs to the convex cover of $\mathfrak{S} \cap \operatorname{Fr}\mathfrak{X}$, and thus $\mathfrak{S}$ is the circumsphere of $\mathfrak{X}$ and the theorem is proved.

COROLLARY. *The radius of the circumsphere of a set $\mathfrak{X}$ of constant width λ lies between $\frac{1}{2}\lambda$ and $\lambda[n/(2n+2)]^{\frac{1}{2}}$.*

Both these limits are attained, the first for a sphere and the second for any set of constant width λ which contains a regular simplex of diameter λ. When $n=2$ this second class reduces essentially to one set, the Reuleaux triangle formed from three equal circular arcs of radius λ, but when $n>2$ there are many sets that belong to this class.

The first of these bounds follows from the facts that

$$R(\mathfrak{X}) \geqslant r(\mathfrak{X}), \quad R(\mathfrak{X}) + r(\mathfrak{X}) = \lambda.$$

The second follows from Theorem 49 and the fact that $D(\mathfrak{X}) = \lambda$.

We next prove the fundamental relation referred to in the introduction.

THEOREM 54. *Every set of diameter D is contained in a set of constant width D.*

If $\mathfrak{X}$ is a point set of diameter D, the convex cover of its closure also has the same diameter as $\mathfrak{X}$, and so we may suppose without loss of generality that it is a closed convex set. If $\mathfrak{X}$ is not of constant width it is not complete. We next define a sequence of sets, as follows. Write

$$\mathfrak{A}(\mathfrak{X}) = \{\mathbf{x}; D(\mathbf{x} \cup \mathfrak{X}) = D(\mathfrak{X})\}$$

$$\rho(\mathfrak{X}) = \sup \rho(\mathbf{x}, \mathfrak{X}) \quad \text{where} \quad \mathbf{x} \in \mathfrak{A}(\mathfrak{X})$$

$$\mathfrak{B}(\mathfrak{X}) = \{\mathbf{x}; \mathbf{x} \in \mathfrak{A}(\mathfrak{X}) \quad \text{and} \quad \rho(\mathbf{x}, \mathfrak{X}) = \rho(\mathfrak{X})\}.$$

Then select $\mathbf{x}_1 \in \mathfrak{B}(\mathfrak{X})$ and define $\mathfrak{X}_1 = \mathfrak{H}(\mathbf{x}_1 \cup \mathfrak{X})$. Generally when $\mathfrak{X}_i$ has been defined select $\mathbf{x}_{i+1} \in \mathfrak{B}(\mathfrak{X}_i)$ and define $\mathfrak{X}_{i+1} = \mathfrak{H}(\mathbf{x}_{i+1} \cup \mathfrak{X}_i)$.

The sequence of sets $\{\mathfrak{X}_i\}$ converges to a set $\bar{\mathfrak{X}}$. $\bar{\mathfrak{X}}$ is convex and $D(\bar{\mathfrak{X}}) = D$. We shall show that $\bar{\mathfrak{X}}$ is complete. If $\bar{\mathfrak{X}}$ is not complete let $\mathbf{y}$ be a point such that $\mathbf{y} \notin \bar{\mathfrak{X}}$ and $D(\mathbf{y} \cup \bar{\mathfrak{X}}) = D$. Let $\delta = \rho(\mathbf{y}, \bar{\mathfrak{X}})$. Now consider two points $\mathbf{x}_i, \mathbf{x}_j, j > i$. Since $\mathbf{x}_i \in \mathfrak{X}_i \subset \mathfrak{X}_{j-1}$ we have

$$|\mathbf{x}_j - \mathbf{x}_i| > \rho(\mathfrak{X}_{j-1}).$$

But
$$|\mathbf{y} - \mathbf{x}| \geqslant \delta, \quad \mathbf{x} \in \mathfrak{X}_j,$$

and this implies $\rho(\mathfrak{X}_{j-1}) \geqslant \delta$. But all the points $\{\mathbf{x}_i\}$ lie in a bounded portion of R^n, and it is impossible for every two of them to be at a distance greater than or equal to δ apart.

Hence the theorem is proved.

Miscellaneous remarks. A property of sets of constant width which follows from the fact that every two parallel support hyperplanes are at the same minimal width apart, is that every normal is a double normal, i.e. a line which passes through a frontier point $\mathbf{x}$ of a set of constant width $\mathfrak{X}$ and is perpendicular to a support hyperplane to $\mathfrak{X}$ at $\mathbf{x}$, meets the frontier of $\mathfrak{X}$ again at $\mathbf{y}$ and $\mathbf{xy}$ is perpendicular to a support hyperplane to $\mathfrak{X}$ at $\mathbf{y}$.

Every support hyperplane to $\mathfrak{X}$ is regular because of the identity $\mathfrak{S}(\mathfrak{X}) = \mathfrak{X}$, but there may be more than one support hyperplane through one frontier point of $\mathfrak{X}$.

A convergent sequence of sets of constant width λ converges to a set of constant width λ.

If $\mathfrak{X}_1, \mathfrak{X}_2$ are of constant width λ_1, λ_2, then $\mathfrak{X}_1+\mathfrak{X}_2$ is of constant width $\lambda_1+\lambda_2$.

The projection of a set of constant width λ is an $(n-1)$-dimensional set of constant width λ, and, conversely, if every projection of a set is of constant width so is the set itself.

2. Plane sets of constant width

Much more information has been found concerning plane sets than n-dimensional sets, and we give in this paragraph a number of properties for most of which we do not know whether the n-dimensional analogues are true or not.

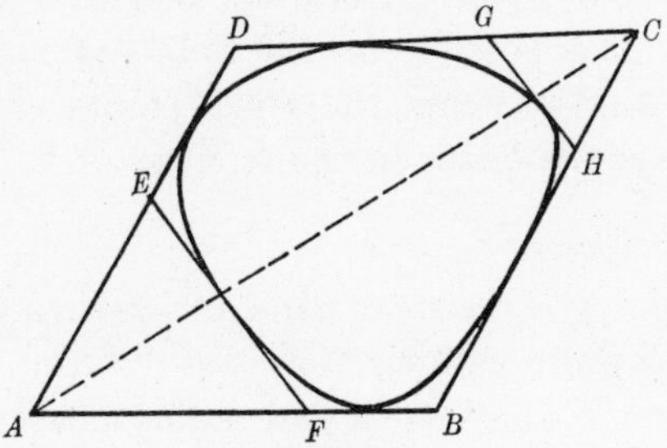

Fig. 14

(i) *The length of the perimeter*

Every plane set $\mathfrak{X}$ of constant width λ has length of perimeter $\pi\lambda$. For by Cauchy's formula § 5·3 this length L is given by

$$L=\frac{1}{2}\int_0^{2\pi} B(\phi)\,d\phi,$$

where $B(\phi)$ is the length of the projection of $\mathfrak{X}$ in the direction ϕ. This formula applied with $B(\phi)=\lambda$ gives $L=\pi\lambda$.

(ii) *Circumscribing hexagons*

An important property of a plane set $\mathfrak{X}$ of constant width λ is that it can be inscribed in a regular hexagon of minimal width λ (the word inscribed means that every side of the hexagon is a support line of $\mathfrak{X}$).

For we may circumscribe a rhombus, say $ABCD$, about $\mathfrak{X}$ such that the angle at A is $\frac{1}{3}\pi$ (see fig. 14). The two support lines perpendicular to the diagonal AC of $ABCD$ cut off two triangles, say AEF and CHG, from $ABCD$. If EF is of the same length as HG, then the hexagon with vertices E, F, B, H, G, D is regular and circumscribes $\mathfrak{X}$. Otherwise, by a continuity argument (see 6·4 (i)) as we rotate $ABCD$ about $\mathfrak{X}$ the function $EF-HG$ changes sign and is therefore zero for some appropriate orientation of $ABCD$. This proves the result.

(iii) *Radii of curvature*

If the radius of curvature of the frontier of $\mathfrak{X}$ exists at $\mathbf{x}$ and is κ, then the radius of curvature exists at $\mathbf{y}$, where $\mathbf{y}$ is the other point on the normal to the frontier of $\mathfrak{X}$ at $\mathbf{x}$, and is equal to $\lambda-\kappa$.

This follows immediately from the fact that a normal is a double normal. It is possible for the radius of curvature not to exist either at $\mathbf{x}$ or at $\mathbf{y}$, but the set of points where the radius of curvature exists is dense in the frontier of $\mathfrak{X}$.

(iv) *Reuleaux polygons*

A Reuleaux polygon is a set of constant width λ whose frontier is formed from a finite number of arcs of radius λ. They may be drawn as follows. With centre $\mathbf{a}$ and radius λ describe an arc $\mathbf{bc}$. With centre $\mathbf{b}$ describe an arc $\mathbf{ad}$ such that $|\mathbf{c}-\mathbf{d}|<\lambda$. With centre $\mathbf{d}$ describe an arc $\mathbf{be}$ so that $|\mathbf{e}-\mathbf{a}|<\lambda$ and so on. If the arc lengths are chosen appropriately we obtain a closed figure after a finite number of steps. This figure is a Reuleaux polygon.

Any finite set of points of diameter D is contained in a Reuleaux polygon of constant width D.

Any convex set of constant width D can be approximated to arbitrarily closely by an appropriate Reuleaux polygon.

Any convergent sequence of Reuleaux polygons with at most k arcs converges to a Reuleaux polygon with at most k arcs.

(v) *The Blaschke-Lebesgue theorem*

Of all sets of constant width λ the circle has the largest area, by the isoperimetric inequality. We also have

THEOREM 55. *The set of constant width λ which has the least area is the Reuleaux triangle.*

This is the Blaschke-Lebesgue theorem.

We use the notation $H(\theta)$ for the value of the support function $H(\mathbf{u})$, where $\mathbf{u}$ is a unit vector perpendicular to the direction of θ.

Let $\mathfrak{X}$ be a convex plane set of constant width λ circumscribed by a regular hexagon $ABCDEF$ of width λ, and suppose that the notation is such that the order of $ABCDEF$ is in the counterclockwise sense of rotation about the hexagon (see fig. 15). Take O, the centre of the hexagon, as the origin, and the line through O parallel to AB as the initial line. Let $H(\theta)$ be defined as above and $\rho(\theta)$ be the radius of curvature of the frontier of $\mathfrak{X}$ at the

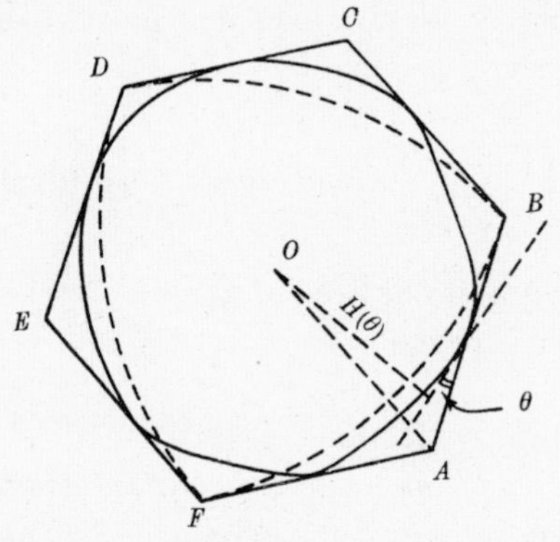

Fig. 15

point where the support line makes an angle θ with the initial direction. In measuring this angle the support line is supposed to be directed in the counterclockwise sense of rotation about $\mathfrak{X}$; $\rho(\theta)$ is zero at a vertex of $\mathfrak{X}$. Similarly, define $H_1(\theta)$ and $\rho_1(\theta)$ with respect to the Reuleaux triangle $\mathfrak{X}_1$, which can be inscribed in $ABCDEF$ to touch it at B, D and F.

Now if two points of $\mathfrak{X}$ lie on a circle S of radius λ then the whole of the smaller arc of S joining these two points also belongs to $\mathfrak{X}$. If we apply this to the arcs BD, DF, FB bounding $\mathfrak{X}_1$ we see that

$$H(\theta) \geqslant H_1(\theta), \quad \tfrac{1}{3}\pi \leqslant \theta \leqslant \tfrac{2}{3}\pi, \quad \pi \leqslant \theta \leqslant \tfrac{4}{3}\pi, \quad \tfrac{5}{3}\pi \leqslant \theta \leqslant 2\pi.$$

Similarly, from a consideration of the Reuleaux triangle obtained by reflecting $\mathfrak{X}_1$ in O it follows that

$$H(\theta+\pi) \geqslant H_1(\theta), \quad \tfrac{1}{3}\pi \leqslant \theta \leqslant \tfrac{2}{3}\pi, \quad \pi \leqslant \theta \leqslant \tfrac{4}{3}\pi, \quad \tfrac{5}{3}\pi \leqslant \theta \leqslant 2\pi.$$

The area of $\mathfrak{X}$ is

$$\frac{1}{2}\int_0^{2\pi}\rho(\theta)\,H(\theta)\,d\theta$$

$$=\frac{1}{2}\left\{\int_{\frac{1}{3}\pi}^{\frac{2}{3}\pi}+\int_{\pi}^{\frac{4}{3}\pi}+\int_{\frac{5}{3}\pi}^{2\pi}\right\}(\rho(\theta)\,H(\theta)+\rho(\theta+\pi)\,H(\theta+\pi))\,d\theta,$$

$$\geqslant\frac{1}{2}\left\{\int_{\frac{1}{3}\pi}^{\frac{2}{3}\pi}+\int_{\pi}^{\frac{4}{3}\pi}+\int_{\frac{5}{3}\pi}^{2\pi}\right\}(\rho(\theta)+\rho(\theta+\pi))\,H_1(\theta)\,d\theta.$$

But $\rho(\theta)+\rho(\theta+\pi)=\lambda$ for all θ except at most an enumerable set. Thus this last integral is equal to

$$\frac{1}{2}\left\{\int_{\frac{1}{3}\pi}^{\frac{2}{3}\pi}+\int_{\pi}^{\frac{4}{3}\pi}+\int_{\frac{5}{3}\pi}^{2\pi}\right\}\lambda H_1(\theta)\,d\theta=\frac{1}{2}\int_0^{2\pi}\rho_1(\theta)\,H_1(\theta)\,d\theta,$$

and this last integral is the area of $\mathfrak{X}_1$.

Thus the result is proved.

(vi) *The asymmetry of convex sets of constant width*

Consider a plane convex set $\mathfrak{X}$ of area $V(\mathfrak{X})$. The largest subset of $\mathfrak{X}$ symmetric with respect to a point $\mathbf{p}$ is written $\mathfrak{X}(\mathbf{p})$. This set is the intersection of $\mathfrak{X}$ with the reflexion of $\mathfrak{X}$ in $\mathbf{p}$. $\mathfrak{X}(\mathbf{p})$ is a central convex set and $\mathbf{p}$ is its centre. The following notations are used.

The coefficient of asymmetry of $\mathfrak{X}$ with respect to $\mathbf{p}$ is

$$f(\mathfrak{X};\,\mathbf{p})=1-V(\mathfrak{X}(\mathbf{p}))/V(\mathfrak{X}).$$

The coefficient of asymmetry of $\mathfrak{X}$ is

$$g(\mathfrak{X})=\min_{\mathbf{p}} f(\mathfrak{X};\,\mathbf{p}).$$

The result of Theorem 51 may be stated as follows. If $\mathfrak{X}$ is any convex set then $g(\mathfrak{X})$ satisfies $0\leqslant g(\mathfrak{X})\leqslant\frac{1}{3}$, and these inequalities are best possible, for the case $g(\mathfrak{X})=0$ holds when $\mathfrak{X}$ is a central set and $g(\mathfrak{X})=\frac{1}{3}$ when $\mathfrak{X}$ is a triangle.

For convex sets of constant width we have the following:

THEOREM 56. *If* $\mathfrak{X}$ *is a convex set of constant width then*

$$0 \leqslant g(\mathfrak{X}) \leqslant g(\mathfrak{R}),$$

where $\mathfrak{R}$ *is a Reuleaux triangle and* $g(\mathfrak{R})$ *is approximately* 0·16.

The proof of this result is omitted; it may be found in the paper by Besicovitch [2].

It is not known what results hold in R^3 analogous to the Blaschke-Lebesgue theorem and Theorem 56.

NOTES

Chapter 1

The definitions and basic properties of convex sets were given by H. Minkowski[1], and a comprehensive account of the subject prior to about 1930 is contained in the book by T. Bonnesen and W. Fenchel[1]. There is also an earlier book by T. Bonnesen[1] and an even earlier book by W. Blaschke, *Kreis und Kugel*. Two recent books are those by H. Hadwiger[1] and by L. Fejes Tóth[1]. The first of these contains a detailed account of the Brünn-Minkowski theory in R^3, whilst the second is particularly concerned with packing and covering problems and a variety of special problems. Axiomatic approaches to convexity have been given by Sperner[1], Stone[1] and Kneser[1].

The theorems in §6 occur in an appendix to *Topologie* by Alexandroff and Hopf. Those of §7 can be proved in many alternative ways. The method used here is essentially that of Bourbaki[1]. See, for example, Brünn[1], Favard[1].

For the properties of cones and their relation to duality the reader should consult Fenchel[3]. An application is given in Karlin and Shapley[1]. See also 'Contributions to the theory of games', *Ann. Math. Study* no. 24, Princeton, N.J. (1950).

Chapter 2

The first proof of Helly's theorem was given by Radon[1] (see also Helly[1], [2], [3]) There have since been many proofs and extensions of it (see, for example, Vincensini[1], [2]). An axiomatic version has been given by Levi[1]. Much of the material in this chapter is contained in Rademacher and Schoenberg[1]. Carathéodory's theorem is given in Carathéodory[1]; see also Hanner and Rådstrom[1]. There is what is essentially an earlier proof of Carathéodory's theorem in Minkowski's *Collected Works*, vol. 2, p. 160. The proof of Kirchberger's theorem is contained in the paper by Rademacher and Schoenberg, *loc. cit.*

For the results of §5 see Horn[1].

Chapter 3

General surveys of the properties of convex functions and their extensions are given in Beckenbach[1] and Green[1]; see also Popoviciu[1]. The use of conjugate functions has been considered by Fenchel[3] and Lorch[1]. The function of Exercise 3·1 (3) was introduced by Mandelbrojt[1]. The criterion for convexity given in §4 is due to Rademacher[1].

See also Bonnesen and Fenchel[1].

CHAPTER 4

For the Blaschke selection theorem see Blaschke[1], Perron[1], Price[1]. The property of distance between neighbourhoods of convex sets which we used was given by Mayer[1]. For the metrization of affinely equivalent classes of convex sets see Reiche[1] and Macbeath[1]. The metrization of the space of support functions has been studied by Süss[1], [2], [3] and Urysohn[1].

CHAPTER 5

For an account of mixed volumes in three dimensions see Hadwiger[1]. See also Vincensini[1] and Bonnesen and Fenchel[1]. For further results the reader should consult the original papers of W. Blaschke, G. Bol, A. Dinghas, E. Schmidt, H. Hadwiger, J. Favard and D. Ohmann.

For the Brünn-Minkowski theorem for general sets see Lusternik[1] and Henstock and Macbeath[1].

For the Fenchel-Alexandroff inequality see Fenchel[1], [2] and Alexandroff[1].

The existence of certain set functions is considered by Fenchel and Jessen[1].

CHAPTER 6

There have been many contributions to the study of special problems. Accounts of some of them are given in the books Bonnesen[1], Fejes Tóth[1] and Pólya and Szëgo[1]. For other results the reader must consult the original works of a very large number of authors, prominent among whom are L. Fejes Tóth, H. Hadwiger and L. Santaló. Earlier contributions were made by W. Süss, and the Japanese school of mathematics, including Fujiwara, Kakeya, Kubota and Matsumura.

The second result of § 1 is due to Bieberbach[1], and Feller's solution of this problem is given in Feller[1]. For the results of § 2 see Bonnesen[1], and for those of § 3 the referee's account of Verblunsky's paper[1] in *Mathematical Reviews*; see also Steinhagen[1] and Gericke[1]. The proof of the corresponding result on spheres is given in Santaló[1]. § 4 is obtained from Besicovitch[1] and Eggleston[3], [4]. Subsection (iii) of § 4 answers a problem proposed by R. C. and E. F. Buck[1]; see also M. Sholander[1].

CHAPTER 7

Considering the number of papers that have been published on sets of constant width it is surprising how little is known about them. The property that such a set can be inscribed in a hexagon is due to Pàl[1]. The Blaschke-Lebesgue theorem was first given independently by Blaschke[2], [3], [4] and Lebesgue[1]. The proof given here is due to Eggleston[2]. The result on asymmetry was first proved by Besicovitch[2]. Further results on sets of constant width are given in Eggleston[1], [5].

REFERENCES

ALEXANDROFF, A. D. [1]. *Rec. Math., Moscow*, **2** (1937), 947–72, 1205–38; **3** (1938), 27–46, 227–51.

BECKENBACH, E. F. [1]. *Bull. Amer. Math. Soc.* **54** (1949), 439–60.

BESICOVITCH, A. S. [1]. *J. Lond. Math. Soc.* **23** (1948), 237–40.

[2]. *Ibid.* 26 (1951), 81–93.

BIEBERBACH, L. [1]. *Jber. dtsch. MatVer.* **24** (1915), 247–50.

BLASCHKE, W. [1]. *Jber. dtsch. MatVer.* **24** (1915), 195–207.

[2]. *Ber. Sächs. Ges. (Akad.) Wiss.*, Math.-Phys. Kl., **66** (1914), 171–7.

[3]. *Ibid.* **67** (1915), 290–7.

[4]. *Math. Ann.* **76** (1915), 504–13.

BONNESEN, T. [1]. *Les problèmes des isopérimetres et des isépiphanes.* Paris (1929).

BONNESEN, T. and FENCHEL, W. [1]. *Theorie der konvexen Körper.* Berlin (1934).

BOURBAKI, N. [1]. *Espaces rectoriels topologiques. Actualités Sci. Industr. nos.* 1189, 1229. Paris (1953) and (1955).

BRÜNN, H. [1]. *Math. Ann.* **100** (1928), 634–7.

BUCK, E. F. and BUCK, R. C. [1]. *Math. Mag.* **22** (1949), 195–8.

CARATHÉODORY, C. [1]. *R. Circ. Mat. Palermo*, **32** (1911), 193–217.

EGGLESTON, H. G. [1]. *Quart. J. Math.* **3** (1952), 63–72.

[2]. *Ibid.* **3** (1952), 296–7.

[3]. *J. Lond. Math. Soc.* **28** (1953), 32–6.

[4]. *Ibid.* **28** (1953), 36–46.

[5]. *Mathematika*, **2** (1955), 48–55.

FAVARD, J. [1]. *J. Math. pures appl.* (9), **12** (1933), 219–82.

FEJES TÓTH, L. [1]. *Lagerungen in der Ebene, auf der Kugel und im Raum.* Berlin (1953).

FENCHEL, W. [1]. *C.R. Acad. Sci., Paris*, **203** (1936), 647–50, 764–6.

[2]. *9th Congr. Math. Skand.* (1939), pp. 249–72.

[3]. *Convex cones, sets and functions* (notes of lectures given at Princeton University) (1953).

FENCHEL, W. and JESSEN, B. [1]. *Danske Vid. Selsk. Mat.-Fys. Medd.* **16**, *no.* 3 (1938), 1–31.

FELLER, W. [1]. *Duke Math. J.* **9** (1942), 885–92.

GERICKE, H. [1]. *Math. Z.* **40** (1935), 317–20.

GREEN, J. W. [1]. *Amer. Math. Mon.* **61** (1954), 449–54.

HADWIGER, H. [1]. *Altes und Neues über konvexe Körper.* Basel and Stuttgart (1955).

HANNER, O. and RÅDSTROM, H. [1]. *Proc. Amer. Math. Soc.* **2** (1951), 589–93.

HELLY, E. [1]. *Mh. Math. Phys.* **31** (1921), 60–91.

[2]. *Jber. dtsch. MatVer.* **32** (1923), 175–6.

[3]. *Mh. Math. Phys.* **37** (1930), 281–302.

HENSTOCK, R. and MACBEATH, A. M. [1]. *Proc. Lond. Math. Soc.* **3** (1953), 182–94.

HORN, A. [1]. *Bull. Amer. Math. Soc.* **55** (1944), 923–9.

KARLIN, S. and SHAPLEY, L. S. [1]. *Geometry of moment spaces.* Memoirs American Math. Soc. no. 12 (1953).

KNESER, H. [1]. *Arch. der Math.* **3** (1952), 198–206.

LEBESGUE, H. [1]. *J. Math. pures appl.* (8), **4** (1921), 67–96.

LEVI, F. W. [1]. *Arch. der Math.* **4** (1953), 222–4.

LORCH, E. R. [1]. *Trans. Amer. Math. Soc.* **71** (1951), 243–66.

LUSTERNIK, L. [1]. *C.R. Acad. Sci. URSS.* **3** (1935), 55–8.

MACBEATH, A. M. [1]. *Canad. J. Math.* 3 (1951), 54–61.

MANDELBROJT, S. [1]. *C.R. Acad. Sci., Paris,* **209** (1939), 977–8.

MAYER, A. E. [1]. *Compos. Math.* **3** (1936), 469–76.

MINKOWSKI, H. [1]. *Theorie der konvexen Körper, insbesondere Begründung ihres Oberflächenbegriffs.* Collected Works, **2**, 131–229.

NEUMANN, B. H. [1]. *J. Lond. Math. Soc.* **14** (1939), 262–72.

PÀL, J. [1]. *Math. Ann.* **83** (1921), 311–9.

PERRON, O. [1]. *Jber. dtsch. MatVer.* **22** (1913), 140–4.

PÓLYA, G. and SZËGO, G. [1]. *Isoperimetric inequalities in mathematical physics. Ann. Math. Study no.* 27. Princeton, N.J. (1951).

POPOVICIU, T. [1]. *Les fonctions convexes. Actualités Sci. Industr. no.* 992. Paris (1944).

PRICE, G. B. [1]. *Bull. Amer. Math. Soc.* **46** (1940), 278–80.

RADEMACHER, H. [1]. *Math. Z.* **13** (1922), 18–27.

RADEMACHER, H. and SCHOENBERG, I. J. [1]. *Canad. J. Math.* **2** (1950), 245–56.

RADON, J. [1]. *Math. Ann.* **83** (1921), 113–5.

REICHE, E. [1]. *Dtsch. Math.* **6** (1941), 171–7, 452, 565.

ROGERS, C. A. and SHEPHARD, G. C. [1]. *J. Lond. Math. Soc.* **33** (1958), 270–81.

SANTALÓ, L. [1]. *Ann. Math.* (2), **47** (1946), 448–59.

SHOLANDER, M. [1]. *Math. Mag.* **24** (1950), 7–10.

SPERNER, E. [1]. *Abh. Math. Sem. Hansischen Univ.* **16** (1944), 140–54.

STEINHAGEN, P. [1]. *Abh. Math. Sem. Hamburg. Univ.* **1** (1922), 15–26.

STONE, M. H. [1]. *Ann. Mat. pura appl.* (4), **29** (1949), 25–30.

SÜSS, W. [1]. *S.B. Preuss. Akad. Wiss.* (1931), pp. 686–95.

[2]. *Tohoku Math. J.* **35** (1932), 326–8.

[3]. *Math. Ann.* **108** (1933), 143–8.

URYSOHN, P. [1]. *Rec. Math., Moscow,* **31** (1924), 477–86.

VERBLUNSKY, S. [1]. *J. Lond. Math. Soc.* **27** (1952), 505–7.

VINCENSINI, P. [1]. *Corps convexes, Séries linéaires, Domains vectoriels. Mém. Sci. Math. France,* **94** (1938).

[2]. *Bull. Sci. Math.* (2), **59** (1935), 163–74.

INDEX